Math Practice
the Singapore Way

Grade 5

Marshall Cavendish
Education

Copyright © 2012 Marshall Cavendish Corporation

Published by Marshall Cavendish Education

Marshall Cavendish Corp.
99 White Plains Road
Tarrytown, NY 10591
Website: www.marshallcavendish.us/edu

Originally published as Maths Practice Papers Copyright © 2002 Times Media Private Limited, Copyright © 2003, 2010 Marshall Cavendish International (Singapore) Limited

Other Marshall Cavendish Offices:
Marshall Cavendish International (Asia) Private Limited, 1 New Industrial Road, Singapore 536196 · Marshall Cavendish International (Thailand) Co Ltd. 253 Asoke, 12th Flr, Sukhumvit 21 Road, Klongtoey Nua, Wattana, Bangkok 10110, Thailand · Marshall Cavendish (Malaysia) Sdn Bhd, Times Subang, Lot 46, Subang Hi-Tech Industrial Park, Batu Tiga, 40000 Shah Alam, Selangor Darul Ehsan, Malaysia.

Marshall Cavendish is a trademark of Times Publishing Limited

ISBN 978-0-7614-8037-2

Printed in the United States
135642

Introduction

Mathematics the Singapore Way is a term coined to refer to the textbook series used in Singapore schools. Mathematics the Singapore Way focuses on problem solving, given that it is based on a curriculum framework that has mathematical problem solving as its focus. Mathematics the Singapore Way also focuses on thinking, given that the Singapore education system is driven by the *Thinking Schools, Learning Nation* philosophy. Mathematics the Singapore Way is also based on learning theories that provide clear directions on how mathematics is learned and should be taught.

Singapore mathematics textbooks, initial teacher preparation, and the subsequent professional development for teachers are based on helping teachers understand what to teach in mathematics and how to teach it.

While these books are not part of the formal classroom program, they provide selected groups of students with the necessary consolidation of skills. Some students require more opportunities to consolidate their basic concepts and skills, and these books are written with that goal in mind. Such learning materials, if they are prepared consistently with the fundamentals of Mathematics the Singapore Way, help learners review basic concepts through the use of visuals. In some cases, it may be necessary for some learners also to have access to concrete materials. It should be noted that good practice is not a matter merely of random repetition. Learners must be helped through careful scaffolding. Good practice consists of careful variations in the tasks learners are given.

I hope this series of books is able to provide the necessary help for learners who need to be challenged beyond basic concepts and skills.

Yeap Ban Har
Marshall Cavendish Institute

Preface

Math Practice the Singapore Way is a series of five books with exercises that adhere closely to the latest Math syllabus in Singapore.

The topics in this book are carefully arranged to follow the sequence of the topics in the students' basic texts, thus enabling them to use the exercises as further practice to gain mastery of mathematical skills. The exercises can also be used to supplement teachers with a thorough and reliable program of testing that provides information for necessary re-teaching.

In mathematics, constant review of concepts is essential. Therefore, the built-in test yourself exercises will refresh students' memories so that no basic skills or concepts will be forgotten.

The variety of sample exam questions found in the test yourself exams at the end of the book will help students face their final math tests with confidence.

Notes pages are provided at the back of the book for those who require more space to work out solutions to the problems.

Contents

Concept: Reading and writing numbers in numerals and in words.

The place value of a digit refers to the value of the digit according to the "place" the number is in.

In 6,732,479,

the digit 9 is in the ones place.
the digit 7 is in the tens place.
the digit 4 is in the hundreds place.
the digit 2 is in the thousands place.
the digit 3 is in the ten thousands place.
the digit 7 is in the hundred thousands place.
the digit 6 is in the millions place.

Millions	Hundred Thousands	Ten Thousands	Thousands	Hundreds	Tens	Ones
6	7	3	2	4	7	9
It stands for 6,000,000	It stands for 700,000	It stands for 30,000	It stands for 2,000	It stands for 400	It stands for 70	It stands for 9

6,732,479 is six million seven hundred thirty-two thousand four hundred and seventy-nine in words.

6,732,479 = 6,000,000 + 700,000 + 30,000 + 2,000 + 400 + 70 + 9

6,732,479 = 6 millions 7 hundred thousands 3 ten thousands 2 thousands 4 hundreds 7 tens 9 ones.

Concept: Comparing and ordering numbers within 10 million.

Work from left to right when comparing or ordering numbers.

1. Which number is larger? Which number is smaller?

Millions	Hundred Thousands	Ten Thousands	Thousands	Hundreds	Tens	Ones
2	4	6	8	2	7	1
2	4	7	4	5	9	4

Compare the digits in the millions place. Both digits are the same. Then, compare the digits in the hundred thousands place. Both digits are the same. Now, compare the digits in the ten thousands place.

7 ten thousands is larger than 6 ten thousands.
So, 2,474,594 is larger than 2,468,271.
6 ten thousands is smaller than 7 ten thousands.
So, 2,468,271 is smaller than 2,474,594.

Math Practice the Singapore Way

Numbers to Ten Million

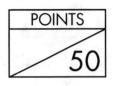

POINTS

50

Section A

Questions 1 through 14 are worth 2 points each. For each question, four options are given. One of them is the correct answer. Make your choice (1, 2, 3, or 4). Write the number of the correct answer in the brackets provided.

1. What is the value of the digit 2 in 235,866?
   (1) 200 thousands   (2) 20 thousands
   (3) 2 thousands   (4) 2 hundreds   ( )

2. In 205,763, the digit 0 is in the _____.
   (1) ten thousands place   (2) thousands place
   (3) hundreds place   (4) units place   ( )

3. What does the digit 6 stand for in 638,524?
   (1) 600   (2) 6,000
   (3) 60,000   (4) 600,000   ( )

4. In 672,586, the 672 tells how many _____ are in it.
   (1) tens   (2) hundreds
   (3) thousands   (4) millions   ( )

5. Write one hundred sixty thousand and thirteen in numerals.
   (1) 106,013   (2) 160,013
   (3) 160,030   (4) 160,130   ( )

6. Write 10,000,000 in words.
   (1) Ten thousand   (2) One hundred thousand
   (3) Ten hundred thousand   (4) Ten million   ( )

7. 5,000 is equal to _____ tens.
   (1) 5   (2) 50
   (3) 500   (4) 5,000   (   )

8. Which one of the following is the largest?
   (1) 109,584   (2) 97,638
   (3) 85,217   (4) 135,218   (   )

9. 500,000 + 8,000 + 10 + 7 = _____
   (1) 5,817   (2) 58,017
   (3) 508,017   (4) 580,017   (   )

10. 6 million and 70 thousand written in numerals is _____.
    (1) 6,000,070   (2) 6,007,000
    (3) 6,070,000   (4) 6,700,000   (   )

11. Three hundred thousand less than 3,333,333 is _____.
    (1) 333,333   (2) 3,033,333
    (3) 3,303,333   (4) 3,333,033   (   )

12. In 377,207 = 300,000 + ☐ + 7,000 + 200 + 7, the missing number
    is _____.
    (1) 700   (2) 7,000
    (3) 70,000   (4) 700,000   (   )

13. 6,208,030 = _____
    (1) 600,000 + 2,000 + 8,000 + 30
    (2) 600,000 + 20,000 + 800 + 30
    (3) 6,000,000 + 200,000 + 800 + 30
    (4) 6,000,000 + 200,000 + 8,000 + 30   (   )

14. What number is ten thousand less than ten million?
    (1) 9,999,000   (2) 9,990,000
    (3) 9,900,000   (4) 9,009,000   (   )

Math Practice the Singapore Way

Section B

Questions 15 through 25 are worth 2 points each. Write your answers in the spaces provided. For questions that require units, give your answers in the units stated.

15. Fifteen thousand and eighty written in numerals is _____.

16. $286{,}050 = 200{,}000 + \boxed{} + 6{,}000 + 50$

    The missing number in the box is _____.

17. Complete the following number pattern:

    $26{,}800, 27{,}800, 28{,}800,$ _____, _____

18. 1,000 more than 47,934 is _____.

19. _____ is 100 less than 20,000.

20. In $28{,}860 = 28{,}760 + \boxed{}$, the missing number in the box is _____.

21. Write 2,874, 900, 31,548, and 2,888 in order, beginning with the largest:

    _____, _____, _____, _____

22. Write 4,160,012 in words.

    _____

23. Which one of the following is the largest?

    34,857, 33,989, 34,875, 34,785 _____

24. Arrange these numbers in decreasing order:
    2,490, 2,049, 4,290, 2,940

    _____, _____, _____, _____

25. Six million and twelve thousand written in numerals is _____.

UNIT 2 Notes

Concept: Rounding off numbers to the nearest thousand.

To round off a number to the nearest thousand, examine the digit in the hundreds place.

We round up if the value of the digit is more than or equal to 5.

We round down if the value of the digit is less than 5.

We can use the approximation sign ≈ which means "approximately equal to."

1. Round off 3,829 to the nearest thousand.
   3,829
   ↓

   8 is more than 5. This means that 3,829 is nearer to 4,000 than 3,000.

So, 3,829 is 4,000 when rounded off to the nearest thousand.

3,829 ≈ 4,000.

2. Round off 284,348 to the nearest thousand.
   284,348
   ↓

   3 is less than 5. This means that 284,348 is nearer to 284,000 than 285,000.

So, 284,348 is 284,000 when rounded off to the nearest thousand.

284,348 ≈ 284,000.

Concept: Estimation.

We can estimate by finding an approximation of a sum or a difference of numbers.

1. Estimate the value of 2,492 + 4,823.

2,492 ≈ 2,000

4,823 ≈ 5,000

2,492 + 4,823 ≈ 2,000 + 5,000

   = 7,000

2. Estimate the value of 5,617 − 3,163.

5,617 ≈ 6,000

3,163 ≈ 3,000

5,617 − 3,163 ≈ 6,000 − 3,000

   = 3,000

3. Estimate the value of 5,938 x 5

5,938 ≈ 6,000

5,938 × 5 ≈ 6,000 × 5

   = 30,000

4. Estimate the value of 6,823 ÷ 5

6,823 ≈ 7,000

6,823 ÷ 5 ≈ 7,000 ÷ 5

   = 1,400

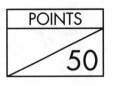

Section A

Questions 1 through 5 are worth 4 points each. For each question, four options are given. One of them is the correct answer. Make your choice (1, 2, 3, or 4). Write the number of the correct answer in the brackets provided.

1. Round off 4,998 to the nearest ten.
    (1) 4,900   (2) 4,990
    (3) 5,000   (4) 5,008   ( )

2. Round off 703,500 to the nearest thousand.
    (1) 700,000   (2) 703,000
    (3) 705,000   (4) 704,000   ( )

3. A number rounded off to the nearest thousand becomes 81,000.
    Which one of the following is the number?
    (1) 79,850   (2) 80,395
    (3) 80,600   (4) 81,525   ( )

4. A number rounded off to the nearest hundred becomes 432,600.
    Which one of the following is the number?
    (1) 432,675   (2) 432,639
    (3) 432,489   (4) 432,540   ( )

5. There are four points on the number line below labeled A, B, C, and D.
    At which point will you mark the number 3,800?

    ```
              A         B         C         D
    ◄──┼──────┼─────────┼─────────┼─────────┼──────►
      3,000                                      4,000
    ```

    (1) A   (2) B
    (3) C   (4) D   ( )

Section B

Questions 6 through 11 are worth 5 points each. Write your answers in the spaces provided. For questions that require units, give your answers in the units stated.

6. How many zeros will be in 2,399,542 when it is rounded off to the nearest

   thousand? _____

7. 5,963 miles is _____ miles when rounded off to the nearest hundred
   miles.

8. Ms. Wilson's income was $37,960 last year. Round off her income to the

   nearest $100. $_____

9. Roger spent $184,340 to renovate his house. Round off the amount to the

   nearest $100. $_____

10. An electronics company sold $106,750 worth of televisions and digital cameras

    during its one-day anniversary sale. Round off the amount to the nearest

    $1,000. $_____

11. 1,712 men, 5,135 women, and 3,264 children attended a holiday fair.
    Round off each number to the nearest thousand by filling in the blanks below.
    Then find the total.

$$
\begin{array}{rcl}
1,712 & \rightarrow & \rule{2cm}{0.4pt} \\
5,135 & \rightarrow & \rule{2cm}{0.4pt} \\
+\ \ 3,264 & \rightarrow & +\ \rule{2cm}{0.4pt} \\
\hline
10,111 & & \rule{2cm}{0.4pt}
\end{array}
$$

Math Practice the Singapore Way

UNIT 3 Notes

Concept: When a number is multiplied by tens, hundreds, or thousands, each digit of the number shifts by one, two, or three places to the place value on its left.

Millions	Hundred Thousands	Ten Thousands	Thousands	Hundreds	Tens	Ones	
			3	8	3	4	
		3	8	3	4	0	x 10
	3	8	3	4	0	0	x 100
3	8	3	4	0	0	0	x 1,000

Concept: When a number is divided by tens, hundreds, or thousands, each digit of the number shifts by one, two, or three places to the place value on its right.

Millions	Hundred Thousands	Ten Thousands	Thousands	Hundreds	Tens	Ones	
2	4	6	1	0	0	0	
	2	4	6	1	0	0	÷ 10
		2	4	6	1	0	÷ 100
			2	4	6	1	÷ 1,000

Concept: Combined operations involving the four operations.

In an expression that involves addition and subtraction, we carry out the operations from left to right.

$3,465 + 6,812 - 5,412$
$= 10,277 - 5,412$
$= 4,865$

In an expression that involves multiplication and division, we also carry out the operations from left to right.

$816 ÷ 4 × 7$
$= 204 × 7$
$= 1,428$

In an expression that involves the four operations of addition, subtraction, multiplication, and division, we carry out the operations from left to right starting with multiplication and division, before addition and subtraction.

$186 + 78 ÷ 6 - 12 × 3$
$= 186 + 13 - 36$
$= 199 - 36$
$= 163$

In an expression that involves brackets, we carry out the operations in the brackets first. After the operations in the brackets, we carry out the operations from left to right starting with multiplication and division, before addition and subtraction.

$(286 - 54 × 3) - 96 ÷ 12$
$= (286 - 162) - 8$
$= 124 - 8$
$= 116$

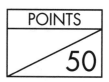
Section A

Questions 1 through 11 are worth 1 point each. For each question, four options are given. One of them is the correct answer. Make your choice (1, 2, 3, or 4). Write the number of the correct answer in the brackets provided.

1. The sum of 528,734 and 3,849 is _____.
   (1) 3,091   (2) 5,101
   (3) 5,111   (4) 16,469   ( )

2. Find the product of 240 and 1,000.
   (1) 1,240   (2) 2,400
   (3) 24,000   (4) 240,000   ( )

3. Which one of the following is equal to 54×49?
   (1) 2,646   (2) 2,546
   (3) 2,516   (4) 702   ( )

4. Tony bought 300 treats for his dogs, Blackie and Brownie. Each day, he gives Blackie 5 treats and Brownie 7 treats. How many days will the treats last?
   (1) 12   (2) 25
   (3) 60   (4) 312   ( )

5. Dewi bought some 25¢ stamps. She gave the postal clerk a $20 bill and received $11 in change. How many 25¢ stamps did she buy?
   (1) 36   (2) 40
   (3) 44   (4) 80   ( )

6. 2,098 people were at a soccer game on Saturday. There were 3 times as many people at a different soccer game on Sunday. How many people were at the soccer games on the two days? Give your answer, correct to the nearest ten.
   (1) 8,392   (2) 8,300
   (3) 8,390   (4) 8,400   (   )

7. Willa made 438 granola bars. She put them equally into 8 plastic bags and had 6 left over. How many granola bars were in each plastic bag?
   (1) 14   (2) 54
   (3) 73   (4) 442   (   )

8. A taxi charges $2.40 for the first mile and 40¢ per mile for the rest of the ride. How much will a 28-mile trip cost?
   (1) $10.80   (2) $11.20
   (3) $13.20   (4) $13.60   (   )

9. Walt saves 50¢ every 3 days. How long will it take him to save $10?
   (1) 6 days   (2) 13 days
   (3) 30 days   (4) 60 days   (   )

10. 4 people had to pay $7.20 to swim in a swimming pool. How much would a group of 6 people have to pay?
    (1) $1.20   (2) $1.80
    (3) $10.80   (4) $43.20   (   )

11. The cost of tiling the floor of a room is $9 per square foot. How much will it cost to tile the floor of a room measuring 16 feet by 14 feet?
    (1) $224   (2) $270
    (3) $540   (4) $2,016   (   )

Section B

Questions 12 through 25 are worth 1 point each. Write your answers in the spaces provided. For questions that require units, give your answers in the units stated.

12. $1{,}586 \times 37 = $ _____

13. $72{,}000 \div 1{,}000 = $ _____

14. $2{,}548 \div 26 = $ _____

15. $120 - 40 \div 5 \times 4 = $ _____

16. $48 + (20 - 8) \div 6 = $ _____

17. $5 \times 6 - 24 \div 6 = $ _____

18. Multiply 3,976 by 9. Then, round off 3,976 to the nearest thousand and multiply it by 9.

$$
\begin{array}{r}
3{,}976 \\
\times \quad 9 \\
\hline
\end{array}
\qquad
3{,}976 \;\rightarrow\; \underline{\qquad} \\
\qquad\qquad \times \; \underline{\qquad} \\
\qquad\qquad\quad \underline{\qquad}
$$

19. Subtract 9,374 from 15,963. Then, round off the numbers below to the nearest thousand before doing the subtraction.

$$
\begin{array}{r}
15{,}963 \;\rightarrow\; \underline{\qquad} \\
- \quad 9{,}374 \;\rightarrow\; - \; \underline{\qquad} \\
\hline
\qquad\qquad\quad \underline{\qquad}
\end{array}
$$

20. 3 apples cost $1. How much will 24 apples cost? $_____

21. 8 cans of corned beef cost $30.80. What is the price of 1 can of corned beef?

    $_____

Math Practice the Singapore Way

22. A parking garage charges $3 for the first hour and $2.50 for every extra hour or part of an hour. How much must Lloyd pay to park his car for 2 hours and 50 minutes?

   $_____

23. A receptionist types 3,500 words in 5 minutes.
   She types _____ words per minute.

24. The water from a faucet flows into a tank at 50 quarts per minute. What is the volume of water in the tank after 1 hour? _____ quarts

25. The postage for sending a package by air to Australia is $17 for the first 250 ounces and $2 for every additional 250 ounces or part of an ounce. Find the postage for sending a 700-ounce package by air to Australia. $_____

Section C

Calculators are allowed in this section.

For questions 26 through 33, show your work clearly in the space below each question and write your answers in the spaces provided. The number of points you can earn is shown in brackets [] at the end of each question.

26. Ben has 30 marbles. Bob has half as many marbles as Ben. Bill has twice as many marbles as Ben. How many marbles must Bill give to each of them so that all three will have the same number of marbles? [4 points]

27. Walt bought 18 cases of soda for a party. There were 24 cans in each case. When the party was over, he had 38 cans left. How many cans of soda did people drink during the party? [3 points]

_____

28. Fran has 1,330 beads, and her sister has 1,026 beads. They want to make necklaces of 38 beads each. How many more necklaces can Fran make than her sister? [3 points]

_____

29. Alan bought a computer for $2,200. He made a down payment of $1,180. He paid the balance in 12 monthly installments. How much did he pay each month? [3 points]

_____

30. A printing press can print 4,800 pages of a magazine in 1 hour. How many pages can it print per minute? [3 points]

_____

31. The table shows the costs of sending a package to Canada by air.

Weight	Charges
First 250 grams	$20
Every additional 250 grams or part of that number	$3.50

Mr. Smith sent a package weighing 900 grams to a friend in Canada. How much did he pay? [3 points]

_____

32. Eunice saves 90¢ per week. She has saved $5.40. How many more weeks must she save to buy a tennis racket that costs $18? [3 points]

_____

33. A sewing machine makes 1,120 stitches for every 80 centimeters. How many stitches will it make for a length of 1.3 meters? [3 points]

_____

Assess Yourself 1

Section A

Questions 1 through 8 are worth 2 points each. For each question, four options are given. One of them is the correct answer. Make your choice (1, 2, 3, or 4). Write the number of the correct answer in the brackets provided.

1. Which one of the following is the largest?
    (1) One million
    (2) 189,725
    (3) 900,000 + 8,000 + 20
    (4) Nineteen thousand and ninety-eight   (   )

2. Which one of the following is **not** equal to 14,213?
    (1) Fourteen thousand two hundred and thirteen
    (2) 14,000 + 213
    (3) 100,000 + 4,000 + 200 + 10 + 3
    (4) $(1 \times 10{,}000) + (4 \times 1{,}000) + (2 \times 100) + (1 \times 10) + (3 \times 1)$ (   )

3. Which one of the following is the best estimate for 69×151?
    (1) 69×150 (2) 69×160
    (3) 70×150 (4) 70×151 (   )

4. Write 6,954, correct to the nearest hundred.
    (1) 6,000   (2) 6,900
    (3) 6,950   (4) 7,000

5. $306 \times 1{,}000 =$ _____
    (1) 3,060,000   (2) 306,000
    (3) 30,600   (4) 3,060   (   )

6. Write ten million in numerals.
   (1) 1,000
   (2) 10,000
   (3) 100,000
   (4) 10,000,000   (   )

7. Bryan's puppy costs $198 and his kitten costs $65 less than his puppy.
   How much do the pets cost altogether?
   (1) $331
   (2) $263
   (3) $253
   (4) $133   (   )

8. Daniel spent $8 and saved $14 of his wages each day.
   How much did he earn in 5 days?
   (1) $30
   (2) $40
   (3) $70
   (4) $110   (   )

Section B

Questions 9 through 15 are worth 2 points each. Write your answers in the spaces provided. For questions that require units, give your answers in the units stated.

9. What whole number comes after 99,999? _____

10. Write the largest possible five-digit number, using the digits 3, 7, and 2 just once and as many zeros as necessary. _____

11. $(7 \times 10) + (4 \times 100) + (3 \times 100,000) + (9 \times \boxed{}) = 390,470$

    The missing number in the box is _____.

12. Divide 1,377 by 9 and give your answer, correct to the nearest ten_____

13. Write 190,012 in words.

    _____

14. Two numbers differ by 7. The larger number is 15. The smaller number

    is _____.

15. T-shirts are sold in boxes of 12. A store owner pays $715.20 for 4 boxes of

    T-shirts. What is the cost of each T-shirt? $_____

Section C

Calculators are allowed in this section.

For questions 16 through 19, show your work clearly in the
space below each question and write your answers in the spaces
provided. The number of points you can earn is shown in brackets []
at the end of each question.

16. 15 teachers and 14 parents took some students on a field trip. The number of
    boys was 9 more than the number of girls. There were 396 people altogether.
    How many girls were on the trip? [5 points]

    Ans: _____

17. Fred bought three bags of concrete, Bag A, Bag B, and Bag C. The total weight
    of the 3 bags of concrete was 78 pounds. Bag A was twice as heavy as Bag B.
    Bag C was 6 pounds heavier than Bag B. What was the difference in weight
    between Bag A and Bag C? [5 points]

    Ans: _____

18. A man bought a set of living room furniture worth $12,236. The down payment was $3,500, and the balance was paid monthly for 2 years. How much did he pay each month? [5 points]

_____

19. A businessperson shipped 8 containers of handbags to Korea. 4 containers contained 1,428 handbags each. 2 containers contained 795 handbags altogether, and each of the remaining containers contained 243 handbags fewer than those in each of the first 4 containers. How many handbags were shipped altogether? [5 points]

_____

Concept: Association of a fraction with division.

Fractions and division are closely related.

Cut a pie into three equal parts.
Each part will then be one-third of the whole.

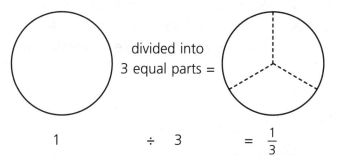

divided into
3 equal parts =

$$1 \qquad \div \qquad 3 \qquad = \frac{1}{3}$$

Concept: Conversion between fractions and decimals.
We can convert a fraction to a decimal by: i) finding its equivalent fraction with a denominator of 10, 100, or 1,000, or ii) doing long division.

Express $\frac{20}{125}$ as a decimal.

$$\frac{20}{125} = \frac{20 \times 8}{125 \times 8}$$
$$= \frac{160}{1,000}$$
$$= 0.16$$

Express $1\frac{3}{8}$ as a decimal.

$$1\frac{3}{8} = 1 + 0.375$$
$$= 1.375$$

$$\begin{array}{r} 0.375 \\ 8\overline{)3.0} \\ \underline{2\ 4} \\ 6\ 0 \\ \underline{5\ 6} \\ 4\ 0 \\ \underline{4\ 0} \\ 0 \end{array}$$

Concept: Addition and subtraction of proper fractions and mixed numbers.

We can add and subtract unlike fractions by converting the fractions into like fractions and adding them together.

Add $\frac{3}{8}$ and $\frac{1}{2}$.

$$\frac{3}{8} + \frac{1}{2} = \frac{3}{8} + \frac{1 \times 4}{2 \times 4}$$
$$= \frac{3}{8} + \frac{4}{8}$$
$$= \frac{7}{8}$$

$$\frac{3}{4} - \frac{7}{12} = \frac{3 \times 3}{4 \times 3} - \frac{7}{12}$$
$$= \frac{9}{12} - \frac{7}{12}$$
$$= \frac{2}{12} = \frac{1}{6}$$

We can add or subtract mixed numbers by first converting the fractional parts of the mixed numbers to like fractions, then adding or subtracting the whole numbers before adding or subtracting the fractional parts.

Add $4\frac{1}{2}$ and $2\frac{5}{12}$.

$$4\frac{1}{2} + 2\frac{5}{12} = 4\frac{6}{12} + 2\frac{5}{12}$$
$$= 6\frac{11}{12}$$

Subtract $6\frac{5}{8}$ and $1\frac{1}{4}$.

$$6\frac{5}{8} - 1\frac{1}{4} = 6\frac{5}{8} - 1\frac{2}{8}$$
$$= 5\frac{3}{8}$$

Concept: Multiplication of a proper fraction and a proper/improper fraction.

We can multiply a fraction by another fraction by first multiplying the numerators and then multiplying the denominators separately.

Multiply $\frac{5}{8}$ by $\frac{2}{3}$.

$$\frac{5}{8} \times \frac{2}{3} = \frac{5 \times 2}{8 \times 3}$$
$$= \frac{10}{24} = \frac{5}{12}$$

Multiply $\frac{7}{4}$ by $\frac{5}{2}$.

$$\frac{7}{4} \times \frac{5}{2} = \frac{7 \times 5}{4 \times 2}$$
$$= \frac{35}{8} = 4\frac{3}{8}$$

Concept: Multiplication of a mixed number and a whole number.

We can multiply a mixed number with a whole number by first converting the mixed number to an improper fraction, and then multiplying the numerator of the improper fraction by the whole number.

Multiply $1\frac{3}{4}$ by 3.

$$1\frac{3}{4} \times 3 = \frac{7}{4} \times 3$$
$$= \frac{21}{4} = 5\frac{1}{4}$$

Concept: Division of a proper fraction by a whole number.
When we divide a fraction by a whole number, we are dividing the fraction into smaller parts.

Divide $\frac{3}{4}$ by 3.

It is the same as finding $\frac{1}{3}$ of $\frac{3}{4}$.
So, we can also say:

$$\frac{3}{4} \div 3 = \frac{3}{4} \times \frac{1}{3}$$
$$= \frac{1}{4}$$

Section A

Questions 1 through 8 are worth 2 points each. For each question, four options are given. One of them is the correct answer. Make your choice (1, 2, 3, or 4). Write the number of the correct answer in the brackets provided.

1. Which one of the following is **incorrect**?

    (1) $\frac{10}{4} = 2\frac{1}{2}$
    (2) $\frac{2}{3} = 2 \div 3$
    (3) $\frac{2}{3} = \frac{4}{9}$
    (4) $3\frac{5}{8} = \frac{29}{8}$ (   )

2. Which one of the following is **incorrect**?

    (1) $\frac{2}{3} + \frac{3}{4} = 1\frac{5}{12}$
    (2) $\frac{3}{4} + \frac{5}{6} = 1\frac{7}{12}$
    (3) $1\frac{1}{4} - 1\frac{7}{8} = 2\frac{1}{8}$
    (4) $1\frac{2}{3} + 2\frac{1}{6} = 3\frac{5}{6}$ (   )

3. Which one of the following is **incorrect**?

    (1) $\frac{5}{8} - \frac{1}{3} = \frac{7}{24}$
    (2) $2\frac{3}{4} - 1\frac{5}{6} = 1\frac{1}{12}$
    (3) $2\frac{3}{4} - 1\frac{5}{8} = 1\frac{1}{8}$
    (4) $5 - 2\frac{1}{4} = 2\frac{3}{4}$ (   )

4. Which one of the following is **incorrect**?

    (1) $\frac{5}{6} \times \frac{9}{10} = \frac{3}{4}$
    (2) $\frac{5}{8} \times \frac{4}{5} = \frac{1}{2}$
    (3) $\frac{9}{10} \div 6 = \frac{3}{20}$
    (4) $\frac{3}{4} \div 3 = 2\frac{1}{4}$ (   )

5. By how many minutes is $3\frac{1}{4}$ hours longer than $1\frac{7}{12}$ hours?

    (1) $60\frac{2}{3}$ minutes
    (2) 62 minutes

    (3) 75 minutes
    (4) 100 minutes   (   )

6. John had $120. He spent $15 in one store and $25 in another. What fraction of his money did he have left?

(1) $\frac{1}{8}$ (2) $\frac{2}{4}$ (3) $\frac{3}{5}$ (4) $\frac{2}{3}$ ( )

7. Paul earns $900 a month. He spends $630 and saves the remainder. What fraction of his expenses are his savings?

(1) $\frac{3}{7}$ (2) $\frac{4}{7}$

(3) $\frac{3}{10}$ (4) $\frac{7}{10}$ ( )

8. Jeff wants to buy a camera that costs $380. He has only $\frac{5}{8}$ of the money needed. How much more does he need?

(1) $76 (2) $142.50

(3) $288 (4) $237.50 ( )

Section B

Questions 9 through 15 are worth 2 points each. Write your answers in the spaces provided. For questions that require units, give your answers in the units stated.

9. Find the value of $\frac{2}{3} \div 4$. _____

10. When a number is divided by 32, the answer is $\frac{5}{8}$. What is the number?

_____

11. How many times is a bag weighing $2\frac{4}{5}$ kilograms as heavy as a bag weighing

400 grams? _____

12. $6\frac{1}{4}$ kg $- 1\frac{1}{2}$ kg $+ 1\frac{2}{5}$ kg $=$ _____ kg _____ g

13. Find the value of $1\frac{2}{3} + 2\frac{3}{5}$. _____

14. Write $\frac{5}{8}, \frac{2}{3}, \frac{5}{6}$, and $\frac{3}{4}$ in order, beginning with the smallest:

_____, _____, _____, _____

15. Find the product of $\frac{3}{10}$ and $\frac{8}{9}$. _____

Calculators are allowed in this section.

For questions 16 through 19, show your work clearly in the space below each question, and write your answers in the spaces provided. The number of points you can earn is shown in brackets [] at the end of each question.

16. Joe spent $\frac{1}{3}$ of his allowance on food and $\frac{3}{4}$ of the rest on transportation. He had $23 left. How much was his allowance? [5 points]

_____

17. Alice had some stamps. She pasted $\frac{1}{4}$ of them on one page, 19 of them on another page, and the remaining 17 stamps on a third page. How many stamps were there altogether? [5 points]

_____

18. A pitcher contained 3 pints of water. Sandra drank $\frac{5}{12}$ of it and Wendy drank $\frac{1}{3}$ of a pint of it. How much water was left in the pitcher? [5 points]

_____

19. Ms. Thompson bought $\frac{3}{4}$ kilograms of raisins. She used $\frac{1}{5}$ of them to bake a cake and $\frac{2}{3}$ of the remainder to bake some cookies. She shared the remaining raisins equally among her two children. How many grams of raisins did each child receive?
[5 points]

_____

Concept: Identifying the base of a triangle and its corresponding height.

Any one side of a triangle may be the base, no matter how it is rotated. The height of a triangle is the perpendicular height from the base to the highest or farthest point of the triangle.

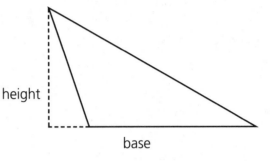

height   base

height   base

We can use mathematical formulas to help us find the area of a triangle.

Area of triangle = $\frac{1}{2}$ × base × height

1. Find the area of the triangle ABC.

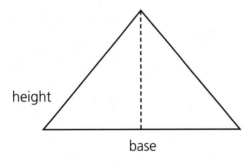

6 cm

C   8 cm   B

A

Area of triangle = $\frac{1}{2}$ × base × height

$\qquad$ = $\frac{1}{2}$ × 8 × 6

$\qquad$ = 24 cm²

2. Find the area of the triangle DEF.

Area of triangle = $\frac{1}{2}$ × base × height

$\qquad$ = $\frac{1}{2}$ × 14 × 5

$\qquad$ = 35 sq. in.

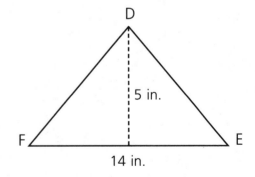

D

5 in.

F   E

14 in.

UNIT 5 — Area of a Triangle

Section A

Questions 1 through 8 are worth 3 points each. For each question, four options are given. One of them is the correct answer. Make your choice (1, 2, 3, or 4). Write the number of the correct answer in the brackets provided.

Study the figures below and answer questions 1 through 4.

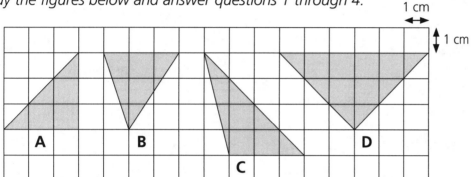

1 cm

1 cm

1. Which of the above triangles has the largest area?
    (1) A   (2) B
    (3) C   (4) D   (   )

2. Which two of the above triangles have the same area?
    (1) A and B   (2) B and C
    (3) A and C   (4) C and D   (   )

3. Which of the above triangles has an area of 6 cm²?
    (1) A   (2) B
    (3) C   (4) D   (   )

4. What is the total area of triangles B and C?

    (1) 10 cm²   (2) $10\frac{1}{2}$ cm²

    (3) $20\frac{1}{2}$ cm²   (4) 21 cm²   (   )

5. Find the area of a triangle whose base is 3.5 inches and height is 5 inches.
   (1) 8.75 square inches   (2) 17.5 square inches
   (3) 35 square inches   (4) 175 square inches   (   )

6. The area of the shaded portion of the figure is _____.
   (1) 11 cm²
   (2) 12 cm²
   (3) 24 cm²
   (4) 36 cm²

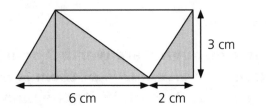

                 (   )

7. The area of the shaded portion of ΔABC is _____.
   (1) 14 cm²
   (2) 28 cm²
   (3) 56 cm²
   (4) 112 cm²

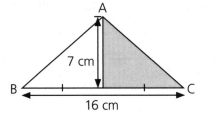

                 (   )

8. The area of the figure is _____.
   (1) 11 cm²
   (2) 14 cm²
   (3) 28 cm²
   (4) 40 cm²

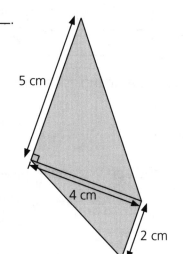

                 (   )

Section B

Questions 9 through 14 are worth 3 points each. Write your answers in the spaces provided. For questions that require units, give your answers in the units stated.

9. ABCD is a rectangle. The measurements are in feet. Find the area of the shaded portion of the rectangle.

    _____ square feet

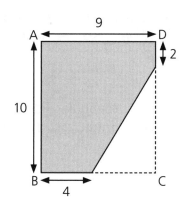

10. Measure the sides of the triangle and find its area. The area of this triangle is _____ square inches.

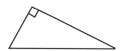

11. Find the area of the figure.

    _____ cm²

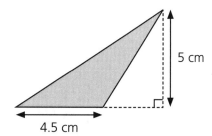

12. Find the area of △ABC.

    _____ cm²

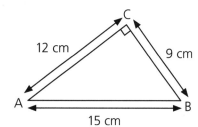

13. A right-angled triangle has sides measuring 6 feet, 8 feet, and 10 feet. What is the area of the triangle? _____ square feet

14. Paul built a pool in the shape of a right-angled triangle with sides that measure 3 yards, 4 yards, and 5 yards. The area of the pool is _____ square yards.

Math Practice the Singapore Way

Section C

 Calculators are allowed in this section.

For questions 15 and 16, show your work clearly in the space below each question, and write your answers in the spaces provided. The number of points you can earn is shown in brackets [] at the end of each question.

15. Each of the four triangular sides of a tent has a base of 5 yards and a height of 4 yards. Find the total area of the four sides of the tent. [4 points]

_____

16. Gary custom-made a lightweight triangular sail for his boat. The base of the sail was 12 yards and the height was 3.2 yards. How much did he have to pay for the canvas for the sail if it costs $118 per square yard? [4 points]

_____

Assess Yourself 2

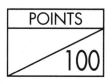
Section A

Questions 1 through 8 are worth 2 points each. For each question, four options are given. One of them is the correct answer. Make your choice (1, 2, 3, or 4). Write the number of the correct answer in the brackets provided.

1. 378,254 when rounded off to the nearest hundred is _____.
    - (1) 378,000
    - (2) 378,200
    - (3) 378,250
    - (4) 378,300   (   )

2. How much more than 1,000 is 1 million?
    - (1) 9,000
    - (2) 9,900
    - (3) 99,000
    - (4) 999,000   (   )

3. Which one of the following is **incorrect**?
    - (1) $\frac{4}{4} = \frac{3}{3}$
    - (2) $3\frac{1}{2} = \frac{7}{2}$
    - (3) $\frac{3}{4} + \frac{1}{3} = \frac{1}{12}$
    - (4) $1\frac{1}{2} - \frac{4}{5} = \frac{7}{10}$ (   )

4. What number when multiplied by 17 is equal to 2,193?
    - (1) 129
    - (2) 2,193
    - (3) 16,779
    - (4) 37,281   (   )

5. Which one of the following is **incorrect**?
    - (1) 52,807 = Fifty-two thousand, eight hundred and seven
    - (2) 65,048 = 6 × 10,000 + 5 × 1,000 + 4 × 10 + 8 × 1
    - (3) 3,864 = 30,000 + 800 + 60 + 4
    - (4) 12 thousands = 10,000 + 2,000   (   )

6. What is the area of a triangle with a base of 32 inches and a height of 16 inches?
   (1) 8 square inches   (2) 16 square inches
   (3) 32 square inches   (4) 256 square inches   (   )

7. There are 18 rows of 15 chairs each in a hall. 237 people are sitting on the chairs. How many empty chairs are there?
   (1) 3   (2) 33   (3) 204   (4) 270   (   )

8. 40 nails are used to make one shoe. How many nails are needed to make 20 similar pairs of shoes?
   (1) 2   (2) 60   (3) 800   (4) 1,600   (   )

Section B

Questions 9 through 15 are worth 2 points each. Write your answers in the spaces provided. For questions that require units, give your answers in the units stated.

9. List all the common factors of 20 and 30. _____

10. $4\frac{2}{3}$ hours = _____ hours _____ minutes

11. Write 80 centimeters as a fraction of 2 meters. _____

12. Find the value of $30 + 6 \div 3 \times 5$. _____

13. Find the area of a rectangle 8 feet long if its perimeter is 24 feet.

    _____ square feet

14. A dealer bought 100 apples for $24 and sold them for 40¢ each.

    How much money did he make? $_____

15. Raul spent $\frac{3}{8}$ of his money in one store and $\frac{1}{6}$ of it in another store.

    He spent $26 altogether. How much did he have at first? $_____

Section C

 Calculators are allowed in this section.

For questions 16 through 19, show your work clearly in the space below each question and write your answers in the spaces provided. The number of points you can earn is shown in brackets [] at the end of each question or part of a question.

16. Jasmine made 250 chicken pot pies. She sold 70 pies in the morning and $\frac{2}{3}$ of the remainder in the afternoon. How many chicken pot pies did she have left?
[5 points]

_____

17. Maggie counted the number of dollar bills in the cash register and found that $\frac{1}{5}$ of the bills were $10 bills and $\frac{1}{2}$ of the remaining number of bills were $5 bills. The rest were made up of 1 $20 bill and 48 $1 bills. What was the total value of the money in the register? [5 points]

_____

18. Kevin bought a high-definition, 3-D television for $4,265 and made a down payment of $350. He paid off the balance in 15 monthly installments. What was the amount of each installment? [5 points]

_____

19. The living room and five bedrooms of a house need new carpeting.
The living room measures 30 feet by 20 feet, while each bedroom has a floor area of 235 square feet.
The carpets cost $12 per square foot. How much will the carpeting cost?
[5 points]

_____

Concept: A ratio is a statement that shows how two or more similar quantities relate to each other in size.

Jenny has some colored beads. 6 are green and 5 are orange.

So, the ratio of the number of green color beads to the number of orange beads is 6 : 5.

Concept: Comparing three quantities.

Susan has some stamps. 24 are local stamps, 8 are foreign stamps, and 10 are collectibles. Express the number of local stamps to the number of foreign stamps to the number of collectibles as a ratio.

24 : 8 : 10
In its simplest form, the ratio is 12 : 4 : 5.

Concept: We can either multiply or divide all the quantities in a ratio to find their equivalent ratios. For example, find an equivalent ratio of 4 : 7.

4 : 7
8 : 14

8 : 14 is an equivalent ratio of 4 : 7.
4 : 7 is the simplest form of 8 : 14.

Concept: We can find one quantity given the other quantity and we can find their ratio.

A length of thread was cut into two pieces in the ratio 3 : 5. The shorter piece was 21 in. Find the length of the longer piece.

21 in.

?

3 units $\longrightarrow$ 21 in.
1 unit $\longrightarrow$ 21 ÷ 3 = 7 in.
5 units $\longrightarrow$ 7 × 5
$\qquad$ = 35 in.

The length of the longer piece is 35 in.

UNIT 6 Ratio

POINTS / 50

Section A

Questions 1 through 8 are worth 2 points each. For each question, four options are given. One of them is the correct answer. Make your choice (1, 2, 3, or 4). Write the number of the correct answer in the brackets provided.

1. Which one of the following is equivalent to 18 : 15?
    (1) 3 : 2   (2) 5 : 6
    (3) 6 : 5   (4) 15 : 18   (   )

2. 30 pounds of prawns are divided into two parts in the ratio 2 : 3. The small and big parts have a weight of _____ respectively.
    (1) 2 pounds and 3 pounds   (2) 6 pounds and 12 pounds
    (3) 12 pounds and 18 pounds   (4) 18 pounds and 12 pounds   (   )

3. The total weight of a box and its contents is 1,875 ounces. The weight of the box is 375 ounces. The ratio of the weight of the box to that of its contents is
    _____.
    (1) 1 : 4   (2) 4 : 1
    (3) 1 : 5   (4) 5 : 1   (   )

4. The lengths of the three sides of a triangle are 30 inches, 40 inches, and 48 inches. The ratio of the length of the shortest side to the length of the longest side is _____.
    (1) 3 : 4   (2) 5 : 8
    (3) 1 : 16   (4) 3 : 48   (   )

5. There are 20 pears, 10 oranges, and 24 apples in a basket. What is the ratio of the number of apples to the total number of pieces of fruit in the basket?
    (1) 4 : 5   (2) 4 : 9
    (3) 5 : 27   (4) 10 : 27   (   )

6. A piece of ribbon is cut into two pieces in the ratio 5 : 3. The whole ribbon is 640 centimeters long. What is the length of the longer piece?

    (1) 80 centimeters   (2) 240 centimeters

    (3) 4 meters   (4) 6 meters   (   )

7. The sides of a triangle are in the ratio 2 : 3 : 4. The longest side is 24 inches. The perimeter of the triangle is _____.

    (1) 18 inches   (2) 54 inches

    (3) 108 inches   (4) 216 inches   (   )

8. The ages of three boys are in the ratio 1 : 3 : 5. Their total age is 57 years. The age of the youngest boy is _____.

    (1) 19 months   (2) 6 years

    (3) 6 years, 4 months   (4) 19 years   (   )

Section B

Questions 9 through 15 are worth 2 points each. Write your answers in the spaces provided. For questions that require units, give your answers in the units stated.

9. Write the ratio 24 : 18 : 12 in its simplest form. _____

10. 1 : 6 = ☐ : 24. The missing number in the box is _____.

11. The length of a rectangle is 18 inches. Its width is 12 inches. The ratio of its length to its perimeter is _____.

12. Mr Long had to pay bills that amounted to $448. He paid $256 to settle some of his bills. Find the ratio of his paid bills to his unpaid bills. _____

13. Measure these lines.

    A ┌─────────────────────────┐ B

          C ┌────────────────┐ D

    The ratio of the length of CD to the length of AB is _____.

         Math Practice the Singapore Way

14. A pole is painted green and red in the ratio 2 : 3. The green portion is 240 centimeters shorter than the red portion. What is the length of the pole in meters? _____ meters

15. 5 : 4 : 2 = 30 : _____ : _____

Section C

 Calculators are allowed in this section.

For questions 16 through 19, show your work clearly in the space below each question, and write your answers in the spaces provided. The number of points you can earn is shown in brackets [] at the end of each question or part of a question.

16. Sharon has $12. Nora has $18. Mary has half as much as Nora. Find the ratio of the amount of Nora's money to the amount of Sharon's money to the amount of Mary's money. [5 points]

_____

17. After Lawrence spent $720 of his salary, he had $180 left. Write the amount he spent as a ratio of his salary in the simplest form. [5 points]

_____

18. The shapes below were formed using wires of the same length. Find the ratio of the area of A to the area of B to the area of C. [5 points]

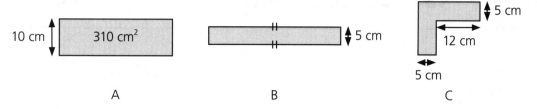

10 cm 310 cm²

A

5 cm

B

5 cm

12 cm

5 cm

C

_____

19. The ratio of the number of boys to the number of girls at a museum was 4 : 15. After 341 girls left and another 136 boys entered the museum, the difference in their number became 612. How many girls were there at first? [5 points]

_____

Assess Yourself 3

Section A

Questions 1 through 8 are worth 2 points each. For each question, four options are given. One of them is the correct answer. Make your choice (1, 2, 3, or 4). Write the number of the correct answer in the brackets provided.

1. In 379,542, the digit 3 stands for _____.
    (1) 3 × 100   (2) 3 × 1,000
    (3) 3 × 10,000   (4) 3 × 100,000   (   )

2. How many 4-mile squares are equivalent in area to a 20-mile square?
    (1) 5   (2) 25
    (3) 80   (4) 100   (   )

3. Beth, Dolly, and Matt shared $144,000 among themselves in the ratio 2 : 3 : 5.
    Matt gave $\frac{1}{8}$ of his share to Beth. How much money did he have left?
    (1) $42,525   (2) $63,000
    (3) $70,875   (4) $81,000   (   )

4. Which one of the following is **incorrect**?
    (1) $2\frac{1}{4}$ kg = 2,250 g   (2) $\frac{3}{8}$ of $\frac{4}{5}$ = $\frac{7}{13}$
    (3) 750 ml = $\frac{3}{4}$ L   (4) $\frac{4}{5} \div 6 = \frac{2}{15}$ (   )

5. Michelle is 12 years older than Steve, who is 8 years old. How old will Michelle be when Steve is 10 years old?
    (1) 18 years old   (2) 20 years old
    (3) 22 years old   (4) 32 years old   (   )

6. 60 students passed and 20 students failed a test. What fraction of the pupils who took the test failed?

(1) $\frac{1}{4}$ (2) $\frac{3}{4}$ (3) $\frac{1}{3}$ (4) $\frac{2}{3}$ ( )

7. Chairs were arranged into 15 rows of 24 chairs each in a school hall. All the chairs except for those in the last two rows were occupied. How many people were seated in the school hall?

(1) 39   (2) 312

(3) 330   (4) 360   ( )

8. Study the figure and answer the following question. The area of the shaded figure is _____.

(1) 17 cm²
(2) 18 cm²
(3) 20 cm²
(4) 22 cm²

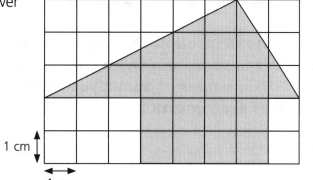

1 cm

1 cm

( )

Section B

Questions 9 through 15 are worth 2 points each. Write your answers in the spaces provided. For questions that require units, give your answers in the units stated.

9. 3,112,002 expressed in words is _____

_____.

10. By what number must 16 be multiplied to make 224? _____

11. The area of a triangle whose base is 13 feet and whose height is 12 feet is

_____ square feet.

12. The product of 47 and 39 rounded off to the nearest hundred is _____.

13. A store owner bought 100 pencils for $27 and sold them at 45¢ each.

    The ratio of the selling price to the cost price of each pencil is _____.

14. A bakery sells buns in bags of 6. How many bags are needed in order for 84

    people to have 2 buns each? _____

15. $(8 \times 3) + 4 \div 2 \times 8 - 4 =$ _____

Section C

 Calculators are allowed in this section.

For questions 16 through 19, show your work clearly in the space below each question, and write your answers in the spaces provided. The number of points you can earn is shown in brackets [] at the end of each question.

16. David ate $\frac{1}{10}$ of a pineapple. He then shared the remainder equally among

    3 people. What fraction of the pineapple did each person receive? [5 points]

17. A florist ordered some flowers from her supplier. The ratio of the number of roses to the number of carnations to the number of lilies is 4 : 7 : 3. By the end of the day, she had sold 84 carnations and 21 roses. The ratio then became 1 : 1 : 1. How many lilies did she order? [5 points]

_____

18. There are 6 boys to every 11 girls in a school. There are 1,254 girls. How many students are there altogether in the school? [5 points]

_____

19. It costs $1,104 to fence a rectangular plot that has an area of 1,080 square yards. The length of the plot is 45 yards. What is the cost of fencing per yard? [5 points]

_____

UNIT 7 Notes

Concept: We can represent a decimal as a fraction by converting it to a fraction with a denominator of 10, 100, or 1,000.

Write the following decimals as fractions or mixed numbers.

a) 1.6

$1.6 = 1 + 0.6$

$= 1 + \dfrac{6}{10}$

$= 1\dfrac{6}{10} = 1\dfrac{3}{5}$

b) 2.51

$2.51 = 2 + 0.51$

$= 2 + \dfrac{51}{100}$

$= 2\dfrac{51}{100}$

c) 8.125

$8.125 = 8 + 0.125$

$= 8 + \dfrac{125}{1000}$

$= 8\dfrac{1}{8}$

Concept: When a decimal is multiplied by tens, hundreds, or thousands, each digit of the number shifts by one, two, or three places to the place value on the left.

Thousand	Hundreds	Tens	Ones	Tenths	Hundreds	Thousandth	
			7	1	3	1	x 10
		7	1	3	1		x 100
	7	1	3	1			x 1,000
7	1	3	1	0			

i. Multiply 7.131 by 10

7 . 1 3 1 × 10 = 71.31

ii. Multiply 7.131 by 100

7 . 1 3 1 × 100 = 713.1

iii. Multiply 7.131 by 1,000

7 . 1 3 1 × 1,000 = 7,131

Concept: When a decimal is divided by tens, hundreds, or thousands, each digit of the number shifts by one, two, or three places to the place value on the right.

Thousand	Hundreds	Tens	Ones	Tenths	Hundreds	Thousandth	
2	4	6	1				÷ 10
	2	4	6	1			÷ 100
		2	4	6	1		÷ 1,000
			2	4	6	1	

i. Divide 2461.0 by 10

2 4 6 1 ÷ 10 = 246.1

ii. Divide 2461.0 by 100

2 4 6 1 ÷ 100 = 24.61

iii. Divide 2461.0 by 1,000

2 4 6 1 ÷ 1,000 = 2.461

UNIT 7 Decimals

POINTS /50

Section A

Questions 1 through 9 are worth 2 points each. For each question, four options are given. One of them is the correct answer. Make your choice (1, 2, 3, or 4). Write the number of the correct answer in the brackets provided.

1. 5 tens 8 hundredths 2 thousandths = _____
    (1) 0.582
    (2) 5.82
    (3) 50.82
    (4) 50.082   (   )

2. In 7.608, the digit 8 is in the _____ place.
    (1) hundredths
    (2) ones
    (3) tenths
    (4) thousandths   (   )

3. Which one of the following is **incorrect**?
    (1) 3,600 g = 3.6 kg
    (2) 2,860 m = 28.6 km
    (3) 5 m, 30 cm = 5.3 m
    (4) 4.3 L = 4 L, 300 ml   (   )

4. Which one of the following is the smallest?
    (1) 0.04
    (2) 0.2
    (3) 1.001
    (4) 0.103   (   )

5. Which one of the following is **incorrect**?
    (1) $0.086 = \frac{8}{100} + \frac{6}{1,000}$
    (2) 0.008 = 8 thousandths
    (3) $0.025 = \frac{1}{40}$
    (4) $0.013 = \frac{13}{100}$ (   )

6. Which one of the following is the best estimate for 59.97 × 30.12?
    (1) 59 × 30
    (2) 60 × 31
    (3) 60 × 30
    (4) 59 × 31   (   )

7. $\frac{3}{8}$ written as a decimal is _____.

    (1) 0.3   (2) 0.375

    (2) 3.8   (4) 8.3   (   )

8. $21,800 \div 1,000 =$ _____

    (1) 2,180   (2) 218

    (3) 21.8   (4) 2.18   (   )

9. $0.007 \times 1,000 =$ _____

    (1) 0.07   (2) 0.7

    (3) 7   (4) 7,000   (   )

Section B

Questions 10 through 17 are worth 2 points each. Write your answers in the spaces provided. For questions that require units, give your answers in the units stated.

10. 2 tens 5 hundredths 8 thousandths written as a decimal is _____.

11. $10\frac{17}{1,000}$ written as a decimal is _____.

12. Arrange these decimals in decreasing order:
2.9, 2.75, 3.1, 3.007

_____, _____, _____, _____

13. 38.005 rounded off to 2 decimal places is _____.

14. 0.075 written as a fraction in the simplest form is _____.

15. Find the difference between 0.02 kilometers and 85 meters. _____ kilometers

16. What fraction of the figure is shaded?

Write your answer as a decimal. _____

17. Write 2.3 as an improper fraction. _____

Section C

For questions 18 through 21, show your work clearly in the space below each question, and write your answers in the spaces provided. The number of points you can earn is shown in brackets [] at the end of each question.

18. A stack of 1,000 sheets of paper is 10 inches high. What is the thickness of each sheet of paper? [4 points]

_____

19. The heights of Mandy, Dominic, Joanne, and Gretel are 0.99 meters, $\frac{4}{5}$ meters, $\frac{7}{8}$ meters, and 1.1 meters, respectively. Order their names according to their heights in decreasing order. [4 points]

_____

20. A cup can hold 275 milliliters of water. It will take 5 cups to fill a pitcher to its brim. The pitcher has 370 milliliters of water at first. How much more water can the pitcher hold? Give your answer in liters. [4 points]

_____

21. Alice placed three strips of balsam wood in a straight line as shown below. Find the total length from A to B. Give your answer in meters. [4 points]

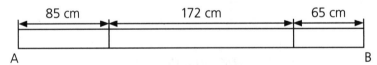

_____

UNIT

8 Notes

Concept: The four operations of decimals.

1. Find 1.23 + 4.15. Express your answer correct to 1 decimal place.
   1.23 + 4.15 = 5.38
   $$\approx 5.4$$

2. Find 6.295 kg – 1.234 kg. Express your answer correct to 2 decimal places.
   6.295 – 1.234 = 5.061
   $$\approx 5.06 \text{ kg}$$

3. Find 4.21 ft. × 4.
   4.21 × 4 = 16.84 ft.

4. Find $37.44 ÷ 8.
   $37.44 ÷ 8 = $4.68

Concept: Converting a measurement from a larger unit to a smaller unit.

Length
1 meter = 100 centimeters
1 kilometer = 1,000 meters
1 foot = 12 inches
1 yard = 3 feet
1 mile = 5,280 feet

Weight
1 kilogram = 1,000 grams
1 pound = 16 ounces

Volume
1 liter = 1,000 milliliters
1 cup = 8 ounces
1 pint = 2 cups
1 quart = 2 pints
1 gallon = 4 quarts

1. Express 3.12 meters in centimeters.
   3.12 × 100 = 312 cm

2. Express 7.15 pounds in ounces.
   7.15 × 16 = 114.4 oz.

3. Express 13.40 gallons in pints.
   13.40 × 4 × 2 = 107.2 pt.

4. Express 4.32 liters in milliliters.
   4.32 × 1,000 = 4,320 ml

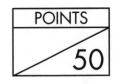
Section A

Questions 1 through 8 are worth 2 points each. For each question, four options are given. One of them is the correct answer. Make your choice (1, 2, 3, or 4). Write the number of the correct answer in the brackets provided.

1. 5 kg, 20 g – 1.57 kg = ☐

    The missing number in the box is _____.
    (1) 3.45 kg   (2) 3 kg, 45 g
    (3) 3.63 kg   (4) 3 kg, 63 g   (   )

2. Find the difference between 3.4 meters and 8 meters 5 centimeters.
    (1) 4.605 meters   (2) 4.65 meters
    (3) 5.1 meters   (4) 11.45 meters   (   )

3. Which one of the following is **incorrect**?
    (1) 0.3 kg × 6 = 1 kg, 800 g
    (2) 1 km, 200 m = 2 × 0.6 km
    (3) 1 m, 80 cm – 0.4 m = 1 m, 40 cm
    (4) 3 L, 20 ml = 16 L ÷ 5   (   )

4. What number is 1 hundredth less than 5.2?
    (1) 4.2   (2) 5.1
    (3) 5.19   (4) 5.199   (   )

5. Jane saves $0.35 a day. How much can she save in July?
    (1) $9.80   (2) $10.05
    (3) $10.50   (4) $10.85   (   )

6. Dan is 1.85 meters tall. Greg is 15 centimeters shorter. When Greg stands on a stool, he is 2.2 meters tall. What is the height of the stool?
    (1) 0.35 meters   (2) 0.5 meters
    (3) 1.7 meters   (4) 2.05 meters   (   )

7. An office cubicle, 7 meters wide, is divided into two smaller cubicles of equal length. What is the length of each smaller cubicle if the area of the original cubicle is 62.3 square meters?
    (1) 4.45 meters   (2) 8.9 meters
    (3) 23.15 meters   (4) 46.3 meters   (   )

8. The curtain rod for each window in Eunice's house measures 175 centimeters. How many meters of curtain rod will she need for 6 such windows in her house?
    (1) 1.05 meters   (2) 10.5 meters
    (3) 105 meters   (4) 1050 meters   (   )

Section B

Questions 9 to 15 are worth 2 points each. Write your answers in the spaces provided. For questions that require units, give your answers in the units stated.

9. $3.26 + 1.48 =$ _____

10. $3.78 \times 38 =$ _____

11. $17.01 \div 9 =$ _____

12. $2.8 \times 1000 =$ _____

13. A bridge is 26 meters long and 2.75 meters wide.
    Its area is _____ square meters.

14. Osman needs 27 boards, each 1 meter, 25 centimeters long.
    Their total length is _____ meters.

15. A watermelon has a weight of 0.75 kilograms. Amy cuts it into 6 equal pieces. The weight of each piece is _____ grams.

Section C

 Calculators are allowed in this section.

For questions 16 to 20, show your work clearly in the space below each question and write your answers in the spaces provided. The number of points you can earn is shown in brackets [] at the end of each question.

16. Mr. Li went to the market on Sunday. He bought 15 pounds of mackeral at $5.65 per pound, 12 pounds of beef at $4.30 per pound, and 6 pounds of prawns at $7.40 per pound. He was then left with $19.25. How much did he bring with him? [4 points]

_____

17. Mr. Peterson wants to fence his ranch using two different types of fencing. Type A fencing costs $253 per kilometer. Type B fencing costs $147 per kilometer. How much will it cost him to fence up the ranch? [4 points]

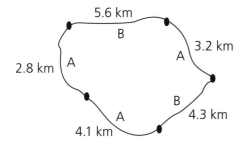

_____

18. A fisher caught 27 pounds of prawns and 21 pounds of trout. He sold his entire catch for $498.75. He sold 1 pound of trout for $5.75. How much did he charge for 1 pound of prawns? [4 points]

_____

Math Practice the Singapore Way

19. Valerie used 1.5 cups of sugar to bake a cake and half as much sugar to make the frosting. She wanted to bake 6 cakes but found that she was short 1.75 cups of sugar. How many cups of sugar did she have? [4 points]

_____

20. A bag of sand weighs 21 pounds. It is divided into three parts. The first part weighs 0.1 pounds more than the second part. The third part weighs 0.7 pounds more than the first part. What does the third part weigh? [4 points]

_____

UNITS

Assess Yourself 4

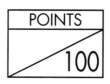
Section A

Questions 1 through 8 are worth 2 points each. For each question, four options are given. One of them is the correct answer. Make your choice (1, 2, 3, or 4). Write the number of the correct answer in the brackets provided.

1. In ☐ : 4 = 18 : 24, the missing number in the box is _____.
   (1) 3   (2) 6
   (3) 14   (4) 72   (   )

2. 5 ones 6 tenths 4 hundredths = _____
   (1) 5.64   (2) 56.04
   (3) 465   (4) 564   (   )

3. Which one of the following is **incorrect**?
   (1) $3 \div 8 = \frac{3}{8}$ (2) $\frac{18}{4} = 4\frac{1}{2}$
   (3) $\frac{2}{3} + \frac{4}{5} = 1\frac{7}{15}$ (4) $1\frac{1}{4} - \frac{1}{3} = \frac{7}{12}$ (   )

4. A bag of dried prunes weighs 400 grams. How many bags of dried prunes will Samuel have if there are 37.2 kilograms of dried prunes?
   (1) 9.3   (2) 93
   (3) 930   (4) 9,300   (   )

5. Which one of the following is **incorrect**?
   (1) $\frac{3}{8} = 0.375$ (2) $2.4 = 2\frac{2}{5}$
   (3) $\frac{12}{2} = 7.6$ (4) $1\frac{3}{1,000} = 1.003$ (   )

6. The base area of a can is 100 square centimeters. Its height is 5 centimeters. Write the capacity of the can as a fraction of 2 liters.

(1) $\frac{1}{4}$

(2) $\frac{1}{10}$

(3) $\frac{1}{20}$

(4) $\frac{1}{250}$

( )

7. Round off 27.348 to 1 decimal place.
   (1) 27.4
   (2) 27.35
   (3) 27.34
   (4) 27.3

( )

8. Jim has $34.50 more than Ellen. They have $80 altogether. How much does Ellen have?
   (1) $22.75
   (2) $45.50
   (3) $57.25
   (4) $114.50

( )

Section B

Questions 9 through 15 are worth 2 points each. Write your answers in the spaces provided. For questions that require units, give your answers in the units stated.

9. Write $\frac{5}{8}$ as a decimal. _____

10. The quotient when 360 is divided by 18 is _____.

11. Write 180 centimeters, 1 meter and 48 centimeters, $1\frac{1}{4}$ meters, and 1.6 meters in order, beginning with the shortest length:

   _____, _____, _____, _____

12. What fraction of the figure is shaded?_____

13. A store owner has 28 boxes, each containing 20 markers. He packs them equally in 4 packages. How many markers are in each package?

   _____

14. $3 \div 1{,}000 =$ _____

15. 1 small bottle holds 125 milliliters of hand soap, and 1 big bottle holds twice as much. 8 small bottles and 4 big bottles hold _____ liters of hand soap.

Section C

 Calculators are allowed in this section.

For questions 16 through 19, show your work clearly in the space below each question, and write your answers in the spaces provided. The number of points you can earn is shown in brackets [] at the end of each question.

16. The area of a square is 196 square inches, and the perimeter of another square is 48 inches. Find the ratio of one side of the big square to one side of the small square. [5 points]

_____

17. Theresa bought $4\frac{1}{2}$ yards of cloth at $4 per yard. Now she is left with only 3.6 yards of cloth. What was the cost of the cloth that she used? [5 points]

_____

18. Joan paid $2.40 for 1 exercise book and 3 pencils. When she bought 1 exercise book and 16 pencils, the store owner charged her $8.25. What was the cost of each item? [5 points]

_____

19. Miguel and Juliana want to buy dinnerware together. A set will cost them $2,800. Miguel has $450 in his savings account, and Juliana has $725 in her savings account. They save $195 per month altogether. How many months will it take them to save enough money to buy all the dinnerware? [5 points]

_____

UNIT 9 Notes

Concept: Using the percentage symbol (%).
We use the symbol "%" to represent percentage.
A percent means 1 out of 100.

Express each of the following as a percentage
1. 0.72
   0.72 × 100 = 72%

ii) $\frac{2}{5}$
   $\frac{2}{5}$ × 100 = 40%

2. Express 15% as a fraction in its simplest form.
   15% = $\frac{15}{100}$
       = $\frac{3}{20}$

3. Express 65% as a decimal.
   65% = $\frac{65}{100}$
       = 0.65

Concept: Expressing a part of a whole as a percentage.

In a bag of marbles, $\frac{3}{8}$ of the marbles are ceramic marbles. What percentage of the marbles are ceramic marbles?

$\frac{3}{8}$ × 100 = 37.5%

37.5% are ceramic marbles.

Concept: Finding a percentage of a whole.

There are 25 books in the class library.
40% of the books are fiction books and the remaining books are nonfiction books.
How many nonfiction books are there?

100% - 40% = 60%

Number of nonfiction books = $\frac{60}{100}$ × 25
                           = 15

There are 15 nonfiction books.

Concept: Solving up to 2-step word problems involving sales tax.

Kelly bought $85 worth of purchases.
If she has to pay 7% in sales tax, how much must she pay in all?

$\frac{7}{100}$ × $85 = $5.95

$85 + $5.95 = $90.95

Kelly has to pay $90.95 in all.

Concept: Solving up to 2-step word problems involving discounts.

Stevie bought a $556 guitar at a 35% discount.
How much did he pay for the guitar?

100% − 35% = 65%

$\frac{65}{100}$ × 556 = $361.40

Stevie paid $361.40 for the guitar.

Concept: Solving up to 2-step word problems involving annual interest.

Jennifer took out a 1-year loan of $25,400 from a bank.
The bank charges 8% interest per year.
How much interest does Jennifer have to pay on the loan?

$\frac{8}{100}$ × $25,400 = $2,032.

Jennifer will have to pay $2,032 in interest on the loan.

Percentage

Section A

Questions 1 through 8 are worth 2 points each. For each question, four options are given. One of them is the correct answer. Make your choice (1, 2, 3, or 4). Write the number of the correct answer in the brackets provided.

1. 12% of $\frac{3}{4}$ kilogram is _____.
   (1) 9 grams   (2) 90 grams
   (3) 108 grams   (4) 9,000 grams   (   )

2. What fraction of 80 is 60% of 50?
   (1) $\frac{3}{8}$ (2) $\frac{5}{8}$
   (3) $\frac{3}{4}$ (4) $\frac{5}{6}$ (   )

3. 12 minutes is _____.
   (1) 12% of 1 hour   (2) 12% of 60 minutes
   (3) 20% of 1 hour   (4) $\frac{1}{12}$ of 1 hour   (   )

4. Which one of the following is the smallest in value?
   (1) $\frac{1}{5}$ of $4   (2) 0.5 of $1.20
   (3) $0.90   (4) 30% of $1.50   (   )

5. A man received a salary of $2,250 a month. He saved $450 of it.
   What percentage of his salary did he spend?
   (1) 5%   (2) 20%
   (3) 50%   (4) 80%   (   )

6. Umbrellas with a list price of $10.80 each are sold at a 25% discount. How much do 6 umbrellas cost with the discount?
   (1) $2.70   (2) $8.10
   (3) $16.20   (4) $48.60   ( )

7. A man bought a computer for $2,800. Sales tax on it cost an extra 7% of the value of the computer. How much did he have to pay for the computer?
   (1) $196   (2) $2,807
   (3) $2,870   (4) $2,996   ( )

8. Out of 40 students in a class, 32 go to school either by bus or by car and the rest walk to school. What percentage of the number of students in the class walk to school?
   (1) 8%   (2) 20%
   (3) 25%   (4) 32%   ( )

Section B

Questions 9 through 15 are worth 2 points each. Write your answers in the spaces provided. For questions that require units, give your answers in the units stated.

9. $\frac{3}{20}$ written as a percentage is _____%.

10. 20% of 1 hour, 20 minutes = _____

11. 9% of 0.4 kilogram = _____ grams

12. What percentage of the figure is **not** shaded? _____%

13. 48% written as a fraction in its lowest terms is _____.

14. Sue gets an allowance of $10 every week. She saves 20% of it. How long will it take her to save $10? _____

     Math Practice the Singapore Way

15. 4 out of 5 students in Ms. Lee's class live in apartments. What percentage of the number of students in Ms. Lee's class do not live in apartments?

_____%

Section C

Calculators are allowed in this section.

For questions 16 through 19, show your work clearly in the space below each question, and write your answers in the spaces provided. The number of points you can earn is shown in brackets [] at the end of each question or part of a question.

16. Marisa had to solve 24 problems. 6 of her solutions were wrong.
    What percentage of the problems did she solve correctly? [5 points]

_____

17. During a sale, a discount of 30% is given on a dress that costs $80.
    Willa has $50.
    (a) How much more money does she need to buy the dress? [3 points]
    (b) What percentage of the money needed for the dress doesn't she have?
    Round off your answer to the nearest percent. [2 points]

(a)_____

(b)_____

18. Karen's weight dropped by 15% in three months. Her weight after the decrease was 38.25 kilograms. What was her weight three months ago? [5 points]

_____

19. A man bought a refrigerator for $1,980. He paid this amount plus 15% interest in monthly installments for 18 months. How much did he pay each month? [5 points]

_____

UNIT 10 Notes

Concept: Making sense of averages.

The average of something is the sum or total amount divided by the number of items.

Average = Sum or Total Amount/Number of Items

Find the average of each of the following sets of items.
i) 5, 6, 8, 3, 6, 8

Sum = 5 + 6 + 8 + 3 + 6 + 8 = 36
Average = 36 ÷ 6 = 6

ii) 2.17 yards, 3.12 yards, 4 yards, 1.39 yards

Sum = 2.17 + 3.12 + 4 + 1.39 = 10.68
Average = 10.68 ÷ 4 = 2.67 yd.

iii) 6.30 kilograms, 23.2 kilograms, 12.50 kilograms

Sum = 6.30 + 23.2 + 12.50 = 42
Average = 42 ÷ 3 = 14 kg

iv) 493 gallons, 539 gallons

Sum = 493 + 539 = 1032
Average = 1032 ÷ 2 = 516 gal.

Concept: Find the total amount given the average and the number of items.

1. The average height of 3 bamboo sticks is 65 centimeters.
   Find the total length of the 3 bamboo sticks.

   65 × 3 = 195

   The total length of the 3 bamboo sticks is 195 centimeters.

2. The average weight of a cookie is 16 ounces.
   Find the total weight of 50 such cookies.

   50 × 16 = 800

   The total weight of 50 such cookies is 800 ounces.

POINTS

/ 50

Section A

Questions 1 through 4 are worth 3 points each. For each question, four options are given. One of them is the correct answer. Make your choice (1, 2, 3, or 4). Write the number of the correct answer in the brackets provided.

1. The average of 5 and 9 is _____.
    (1) 9 + 5   (2) 9 − 5
    (3) 14 × 2   (4) 14 ÷ 2   (   )

2. The average of 25, 24, and 23 is _____.
    (1) 24   (2) 25
    (3) 49   (4) 72   (   )

3. 4 is the average of _____.
    (1) 12 and 2   (2) 3, 6, and 7
    (3) 3 and 5   (4) 3 and 1   (   )

4. The average cost of 3 books is $2.10. Their total cost is _____.
    (1) $0.70   (2) $2.13
    (3) $5.10   (4) $6.30   (   )

Section B

Questions 5 through 10 are worth 3 points each. Write your answers in the spaces provided. For questions that require units, give your answers in the units stated.

5. Pail A can hold 2.7 gallons of water. Pail B and Pail C together can hold

    0.3 gallons more than Pail A. Find the average capacity of the three pails.

    _____ gallons

6. The table shows Trudi's grades, on a 50-point scale, in

   four subjects. Her average grade is _____.

Subjects	Grades
English	43
Math	29
Science	40
Spanish	36

7. Five baskets of vegetables weigh 6.8 pounds, 8.8 pounds, 8.7 pounds,

   11.2 pounds, and 8 pounds. Find their average weight. _____ pounds

8. The lengths of 3 poles are as follows: 3.75 feet, 2.6 feet, and 4.6 feet.

   What is their average length? _____ feet

9. The total weight of nine boys is 407.7 pounds. What is their average weight?

   _____ pounds

10. Ms. Weller weighed six books and found that their average weight
    was 1.75 pounds. One of the books weighed 1.9 pounds.
    Find the total weight of the other five books. _____ pounds

Section C

 Calculators are allowed in this section.

**For questions 11 through 14, show your work clearly in the space
below each question, and write your answers in the spaces
provided. The number of points you can earn is shown in brackets []
at the end of each question.**

11. The average weight of 2 girls is 29 kilograms, 850 grams, and the average
    weight of 3 boys is 32 kilograms, 600 grams. Find their total weight. [5 points]

_____

12. The average age of 4 students is 7 years, 2 months. The average age of 3 of them is 6 years, 9 months. What is the age of the fourth student? [5 points]

_____

13. The total height of 17 students is 23 meters, 25 centimeters. The average height of 4 of them is 1 meter, 75 centimeters. Find the average height of the other 13 students. [5 points]

_____

14. The average weight of 2 packages is 3 kilograms, 725 grams. A third package has a weight of 2 kilograms, 846 grams. What is the average weight of the 3 packages? [5 points]

_____

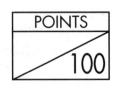
Section A

Questions 1 through 8 are worth 2 points each. For each question, four options are given. One of them is the correct answer. Make your choice (1, 2, 3, or 4). Write the number of the correct answer in the brackets provided.

1. The value of the digit 3 in 163,547 is equal to _____.
   (1) 30×1 (2) 30×10
   (3) 30×100 (4) $30 \times 1,000$ (   )

2. How many times is $7\frac{1}{2}$ kilograms greater than 25 grams?
   (1) 3,000   (2) 300
   (3) 30   (4) 3   (   )

3. Find the value of $16 + 48 \div 8 - 2$.
   (1) 6   (2) 10
   (3) 20   (4) 24   (   )

4. 5.02 kilometers = _____
   (1) 5.2 meters   (2) 52 meters
   (3) 502 meters   (4) 5,020 meters   (   )

5. 48 minutes is _____ of 5 hours.
   (1) 0.096   (2) 0.16
   (3) 0.48   (4) 9.6   (   )

6. In 287.349, which digit is in the thousandths place?
   (1) 9   (2) 4
   (3) 3   (4) 2   (   )

7. How many liters of water can a container measuring 12 centimeters by
    9 centimeters by 10 centimeters hold?
    (1) 0.108 liters   (2) 1.08 liters
    (3) 108 liters   (4) 1,080 liters   (   )

8. Find the area of a triangle with side measurements of 9 inches, 12 inches,
    and 15 inches.
    (1) 108 square inches   (2) 90 square inches
    (3) 67.5 square inches   (4) 54 square inches   (   )

Section B

**Questions 9 through 15 are worth 2 points each. Write your answers
in the spaces provided. For questions that require units, give your
answers in the units stated.**

9. Write eight million and thirty-four thousand in numerals. _____

10. $73.14 \div 7 =$ _____
    (Give your answer, correct to 2 decimal places.)

11. Write 3 liters, 45 milliliters in liters. _____ liters

12. The missing number in $1,000 \times \boxed{} = 600,000$ is _____.

13. Arrange $\frac{2}{3}$, $1\frac{1}{4}$, $\frac{3}{2}$, and $\frac{5}{6}$ in order, beginning with the largest:

    _____ , _____ , _____ , _____

14. Find the value of $1\frac{2}{3} + 2\frac{3}{4}$. _____

15. A rectangular picture, 13 centimeters long
    and 8 centimeters wide, has a 2-centimeter-wide
    margin all around it. The area of the margin

    is _____ square centimeters.

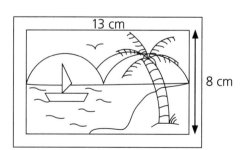

Section C

For questions 16 through 19, show your work clearly in the space below each question, and write your answers in the spaces provided. The number of points you can earn is shown in brackets [] at the end of each question or part of a question.

16. Bag A is twice as heavy as Bag B. The weight of Bag B is 7 kilograms more than the weight of Bag C. The weight of Bag C is 9 kilograms. What is the average weight of the three bags? [5 points]

_____

17. Dresses, originally priced at $240 each, were sold at a discount of 30%. Rita bought a dress at the discounted price. How many scarves, at $12 each, could she buy with the money she saved? [5 points]

_____

18. A sum of money was shared among Henry, Robert, and Alvin in the ratio
    2 : 3 : 5. Robert received $90.
    (a) How much was the sum of money? [2 points]
    (b) What percentage of the money did Henry receive? [3 points]

(a) _____

(b) _____

19. Bill bought a comb for $2.75 and 15 tennis balls at $1.25 each. How much
    change would he receive if he paid for the items using a $50 bill? [5 points]

_____

Assess Yourself 6

Section A

Questions 1 through 8 are worth 2 points each. For each question, four options are given. One of them is the correct answer. Make your choice (1, 2, 3, or 4). Write the number of the correct answer in the brackets provided.

1. 6% written as a fraction is _____.

    (1) 6   (2) $\frac{6}{10}$

    (3) $\frac{2}{3}$ (4) $\frac{6}{100}$ (   )

2. One tenth more than 5.299 is _____.
    (1) 5.3   (2) 5.309
    (3) 5.399   (4) 6.299   (   )

3. 0.2 of one of the following figures is shaded. Which one is it?
    (1)   (2)

    (3)   (4)

                                                           (   )

4. Benny cut a wire, 10.28 feet long, into 8 equal pieces. Find the length of each piece, correct to 2 decimal places.
    (1) 1.3 feet   (2) 1.29 feet
    (3) 1.285 feet   (4) 1.28 feet   (   )

5. Kumar bought a tape recorder for 90 British pounds, including tax. The exchange rate was 1 British pound to $1.40. How much did he have to pay in U.S. dollars?

(1) $92.40   (2) $126.00

(3) $216   (4) $2,160   (   )

6. Ellis bought 3 shirts for $72. The average price of two of the shirts was $19.90. What was the price of the third shirt?

(1) $24   (2) $32.20

(3) $39.80   (4) $52.10   (   )

7. A painter worked from 9:30 a.m. to 5:00 p.m. He was paid $2.50 an hour and spent $2.40 on his lunch. How much did he have left?

(1) $4.30   (2) $7.30

(3) $16.35   (4) $18.75   (   )

8. A square piece of cardboard, which has a perimeter of 12 inches, is cut into two equal triangles. The area of each triangle is _____.

(1) $4\frac{1}{2}$ square inches   (2) 9 square inches

(3) 36 square inches   (4) 72 square inches   (   )

Section B

Questions 9 through 15 are worth 2 points each. Write your answers in the spaces provided. For questions that require units, give your answers in the units stated.

9. Find the total area of the faces of the cube. _____ cm²

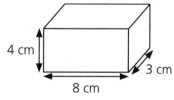

10. Find the value of $5\frac{1}{6} - 2\frac{3}{4}$. _____

11. _____ is 0.9 more than 2.87.

12. Logan earns $684 a month. He saves $\frac{1}{6}$ of this amount each month. How much can he save in a year? $_____

Math Practice the Singapore Way

13. Kara wants to buy mangoes priced at $2.80 for 3 mangoes. How much must she pay for 12 mangoes? $_____

14. A coil of wire is cut into two pieces in the ratio 2 : 3. The shorter piece is 8 feet. What is the length of the longer piece? _____ feet

15. The common factors of the numerator and denominator of $\frac{7}{21}$ are 1 and

_____.

Section C

 Calculators are allowed in this section.

For questions 16 through 19, show your work clearly in the space below each question, and write your answers in the spaces provided. The number of points you can earn is shown in brackets [] at the end of each question.

16. A sum of money is shared among Marcos, Angel, and Rico in the ratio 3 : 2 : 1. Rico receives $78. How much is the sum of money? [5 points]

_____

17. Walter took his family out for dinner. The meal cost $150. He had to pay a service charge of 10% on the meal and another 7% sales tax on the meal and service charge combined. How much did he spend altogether? [5 points]

_____

© 2012 Marshall Cavendish Corporation

18. The table shows the parking charges for a private parking garage.

Hours	Parking charges
1st hour	$3.20
Additional hour or part of hour	$1.45

Mr. Agero parked his car at the parking garage from 9 a.m. to 8:30 p.m. How much did he have to pay in all? [5 points]

_____

19. Jack bought 3 cans of soup, 12 cans of green beans, and 1 can of beef for $39.58. Each can of soup cost $1.90 and the can of beef cost twice as much as one can of green beans. How much did the can of beef cost? [5 points]

_____

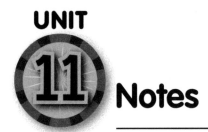

11 Notes

Concept: Angles on a straight line
The sum of angles on a straight line is 180°.

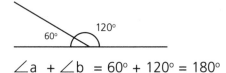

$\angle a + \angle b = 60° + 120° = 180°$

Concept:
Angles at a point
The sum of angles
at a point is 360°.

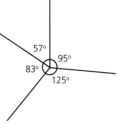

$\angle a + \angle b + \angle c + \angle d = 57° + 83° + 95° + 125°$
$= 360°$

Concept: Vertically opposite angles
Vertically opposite angles are equal.

$\angle a = \angle b = 57°$

Concept: Types of triangles.

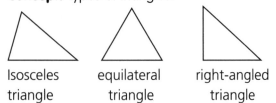

Isosceles   equilateral   right-angled
triangle   triangle   triangle

Concept:
Sum of angles
in a triangle is 180°.

$\angle a + \angle b + \angle c = 48° + 75° + 57° = 180°$

Find the
missing angle.

$\angle a = 180° - 24° - 45° = 111°$

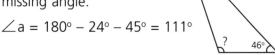

Concept: We can draw a triangle
from given dimensions using ruler,
protractor, and set squares.

Draw a triangle ABC where
i) AB is 4 cm long, and
ii) Angle A is 68°
and Angle B is 57°.

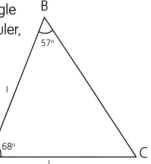

Concept: Parallelogram, rhombus, and trapezium are
four-sided figures.

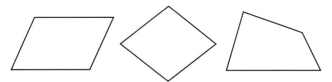

Parallelogram	Rhombus	Trapezium
Opposite sides are parallel.	Opposite sides are parallel.	One pair of opposite sides is parallel.
Opposite sides and opposite angles are equal.	Four sides are equal.	Each pair of angles between parallel sides adds up to 180°.
Each pair of angles between two parallel lines adds up to 180°.	Opposite angles are equal. Each pair of angles between two parallel lines adds up to 180°.	

Concept: We can draw a square / rectangle /
parallelogram / rhombus / trapezium from given
dimensions using ruler, protractor, and set squares.

For example, draw a rhombus where
i) AB is 4 cm long,
ii) AB // DC, and
iii) Angle ABC is 120°.

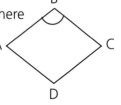

© 2012 Marshall Cavendish Corporation

Angles, Triangles, and 4-sided Figures

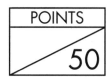

POINTS

/ 50

Figures are not drawn to scale.

Section A

Questions 1 through 11 are worth 2 points each. For each question, four options are given. One of them is the correct answer. Make your choice (1, 2, 3, or 4). Write the number of the correct answer in the brackets provided.

1. AB, CD, and EF are intersecting straight lines.
   $\angle a = 28°$. $\angle b + \angle c = $ _____.
   (1) 28°
   (2) 62°
   (3) 152°
   (4) 332°   (   )

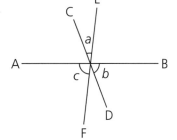

2. Find $\angle x$.
   (1) 67°
   (2) 45°
   (3) 44°
   (4) 22°   (   )

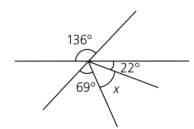

3. Find $\angle k$.
   (1) 110°
   (2) 104°
   (3) 98°
   (4) 68°   (   )

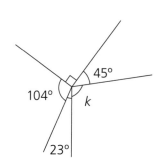

Math Practice the Singapore Way

4. Find the sum of ∠a and ∠b.
   (1) 53°
   (2) 69°
   (3) 110°
   (4) 185°

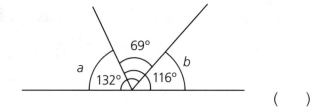
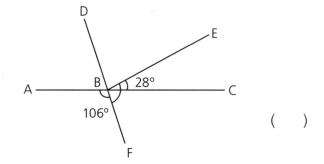

( )

5. Lines ABC, DBF, and BE are straight lines. Find ∠EBF.
   (1) 106°
   (2) 102°
   (3) 78°
   (4) 50°

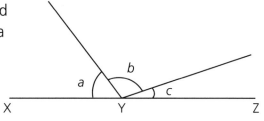

( )

6. ∠a is three times as big as ∠c and ∠b is twice as big as ∠a. XYZ is a straight line. What is the sum of ∠a and ∠c?
   (1) 108°
   (2) 72°
   (3) 54°
   (4) 18°

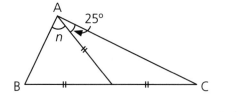

( )

7. ABC is a triangle. The value of n is _____.
   (1) 25°
   (2) 50°
   (3) 65°
   (4) 130°

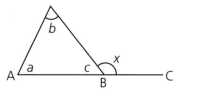

( )

8. ABC is a straight line. Which one of the following statements is **incorrect**?
   (1) ∠a + ∠b = ∠x
   (2) ∠a + ∠b + ∠c = 180°
   (3) ∠b + ∠c = ∠a + ∠x
   (4) ∠c + ∠x = 180°

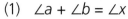
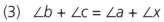
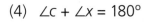

( )

9. $\angle a + \angle b + \angle c = $ _____

    (1) 130°
    (2) 180°
    (3) 310°
    (4) 360°

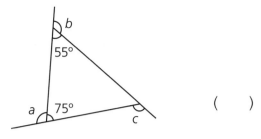

(   )

10. Which one of the following statements about a trapezoid is **incorrect**?
    (1) A trapezoid has two pairs of parallel lines.
    (2) The sum of each pair of angles between two parallel sides is 180°.
    (3) A trapezoid may have two right angles.
    (4) All the angles of a trapezoid may not be equal.   (   )

11. Which one of the following statements describes a rhombus?
    (1) It has one pair of parallel sides.
    (2) One of its angles is 90°.
    (3) It has 4 equal sides, and none of its angles is 90°.
    (4) It has 3 equal sides.   (   )

Section B

Questions 12 through 19 are worth 2 points each. Write your
answers in the spaces provided. For questions that require
units, give your answers in the units stated.

12. $\angle n = $ _____ °

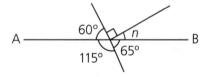

13. $\angle x$ is equal to

    _____ right angles.

In the figure on the right, AB is a straight line.
Use it to answer questions 14 and 15.

14. $\angle b = 40°$. $\angle a = $ _____ °.

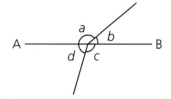

15. $\angle a + \angle b + \angle c = 285°$. $\angle d = $ _____ °.

Math Practice the Singapore Way

16. BAD is a straight line.

    $\angle p =$ _____ °

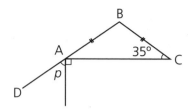

17. QPS and QRT are straight lines.

    The sum of $\angle x$ and $\angle y$ is _____°.

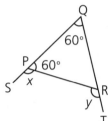

18. ABC is a straight line.

    What is the size of $\angle c$? _____°

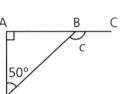

19. ABC is a triangle.

    $\angle n =$ _____°

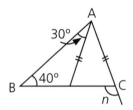

Section C

 Calculators are allowed in this section.

For questions 20 through 25, show your work clearly in the space below each question, and write your answers in the spaces provided. The number of points you can earn is shown in brackets [] at the end of each question.

20. DC // ABE and CB = CE.
    $\angle CEB = 70°$. Find $\angle DCB$. [2 points]

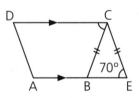

21. ABCD is a rhombus. ∠DCB = 50°.
    Find ∠DBA. [2 points]

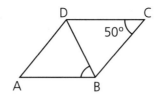

_____

22. ABCD is a rectangle and
    DBEC is a parallelogram.
    Find ∠x. [2 points]

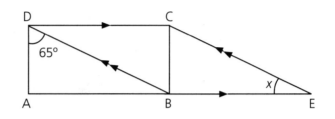

_____

23. Draw a triangle ABC in which AB = 5 centimeters, BC = 6 centimeters, and
    ∠ABC = 80°.
    [2 points]

Math Practice the Singapore Way

24. Draw a parallelogram ABCD in which AB = 5 centimeters, BC = 3 centimeters, and ∠ABC = 110°. Measure ∠BCD. [2 poiints]

_____

25. Draw a rhombus PQRS in which ∠PQR = 125° and PQ = 3.5 centimeters. Measure ∠SPQ. [2 points]

_____

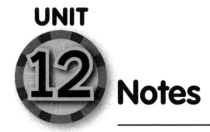
Concept: Building solids with unit cubes.
We may refer to a single cube as a unit cube.
A unit cube typically has 6 faces and 12 equal edges.

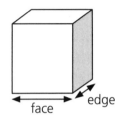

face   edge

When more than one unit cube is placed together to form a solid, we refer to it as a cuboid.

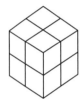

Concept: Measurement of volume in cubic units.
The volume of a unit cube is 1 cubic unit.
We use it to find the volume of a solid made up of unit cubes.

For example, a cuboid is made up of 12 unit cubes. What is the volume of the cuboid?

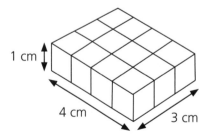

1 cm
4 cm   3 cm

Since each unit cube is 1 cubic unit, 12 unit cubes are 12 cubic units.

Concept: Conversion among liters, milliliters, and cubic centimeters.
The volume of a solid and volume of liquid are related.

1 liter = 1,000 cm³
1 milliliter = 1 cm³

1. Write the following in cubic centimeters.
   i) 234 milliliters   ii) 1,492 milliliters
   Ans.: 234 cm³   Ans.: 1,492 cm³

2. Write the following in litres and milliliters.
   i) 5,843 cm³   ii) 2,006 cm³
   Ans.: 5L, 843 ml   Ans.: 2L, 6 ml

Concept: Using a formula to calculate the volume of a cube/cuboid.
We can use a formula to help us find the volume of cubes/cuboids.
Volume of a cuboid = Length × Width × Height

1. Find the volume of a cuboid by measuring 25 cm by 12 cm by 5 cm.

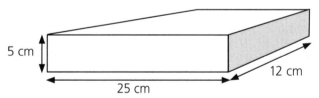

5 cm
25 cm
12 cm

Volume = 25 × 12 × 5 = 1,500 cm³.

2. Find the volume of the cuboid below.

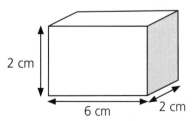

2 cm
6 cm   2 cm

Volume = 6 × 2 × 2 = 24 m³

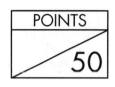

POINTS
/50

Figures are not drawn to scale.

Section A

Questions 1 through 8 are worth 2 points each. For each question, four options are given. One of them is the correct answer. Make your choice (1, 2, 3, or 4). Write the number of the correct answer in the brackets provided.

1. 1 liter of water can fill up a cubic container. The edge of the container is
    _____.
    (1) 1 centimeter   (2) 10 centimeters
    (3) 100 centimeters   (4) 1,000 centimeters   (   )

2. Which one of the following has the greatest volume?
    (1) A container that can hold 1 liter of water
    (2) A cube with a side measurement of 9 centimeters
    (3) A rectangular can measuring 6 centimeters by 6 centimeters by
         5 centimeters
    (4) A box whose volume is 184 cubic centimeters   (   )

3. How many times is the volume of an 8-inch cube as large as the volume of a
    2-inch cube?
    (1) 4 times   (2) 8 times
    (3) 16 times   (4) 64 times   (   )

4. How many cubes with side measurements of 3 feet are equivalent in volume to
    a 12-foot cube?
    (1) 4   (2) 16
    (3) 36   (4) 64   (   )

5. Write 0.07 liters in cubic centimeters.
   (1) 0.7 cubic centimeters   (2) 7 cubic centimeters
   (3) 70 cubic centimeters   (4) 700 cubic centimeters   (   )

6. A rectangular box measures 3 centimeters by 2 centimeters by 4 centimeters. A cube has a volume of 27 cubic centimeters. The ratio of the volume of the cube to the volume of the rectangular box is _____.
   (1) 1 : 3   (2) 3 : 1
   (3) 8 : 9   (4) 9 : 8   (   )

7. A fish tank, 35 centimeters by 20 centimeters by 10 centimeters, is full of water. How many liters of water are in the fish tank?
   (1) 7 liters   (2) 70 liters
   (3) 700 liters   (4) 7,000 liters   (   )

8. The figure shows a stone in a container.

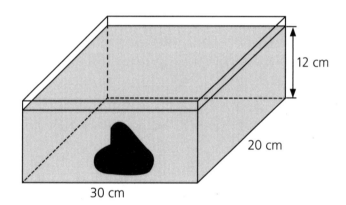

12 cm
20 cm
30 cm

After the stone is taken out, the water level in the container drops from 12 cm to 10 cm. The volume of the stone is _____.
   (1) 1,200 cm³   (2) 3,600 cm³
   (3) 4,800 cm³   (4) 10,800 cm³   (   )

Section B

Questions 9 through 15 are worth 2 points each. Write your answers in the spaces provided. For questions that require units, give your answers in the units stated.

Each of the following solids is made up of 1-centimeter cubes.

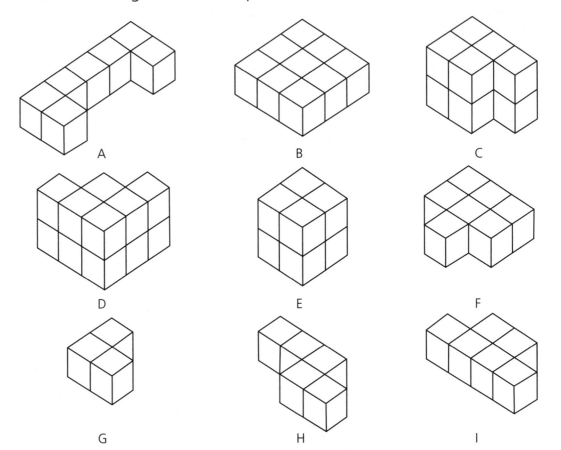

9. The volume of Figure _____ is 8 cubic centimeters.

10. Figure _____ and Figure _____ have the same volume.

11. The volume of Figure I is _____ cubic centimeters.

12. The volume of Figure _____ is twice that of Figure G.

13. The volume of Figure G is $\frac{1}{3}$ of Figure _____.

14. The total volume of Figure D and Figure I is _____ cubic centimeters.

15. The volume of Figure C is greater than Figure I by _____ cubic centimeters.

Section C

 Calculators are allowed in this section.

For questions 16 through 19, show your work clearly in the space below each question, and write your answers in the spaces provided. The number of points you can earn is shown in brackets [] at the end of each question.

16. Draw the cube shown on the dotted paper provided. [5 points]

17. A rectangular water tank is 3 feet long, 2 feet wide, and 4 feet high. It contains 2.7 cubic feet of water. How many more cubic feet of water must be poured into the tank to make it half full? [5 points]

18. A rectangular container, 45 inches long, 48 inches wide, and 20 inches high, was filled to the brim with honey. When Michael took out some honey to fill 36 similar jars, the level of the honey in the container dropped to 16 inches. What was the capacity of each jar? [5 points]

_____

19. A rectangular can measuring 42 centimeters by 28 centimeters by 45 centimeters is full of oil. The oil is then poured into 2-liter bottles. How many bottles are used? [5 points]

_____

Section A

Questions 1 through 15 are worth 2 points each. For each question, four options are given. One of them is the correct answer. Make your choice (1, 2, 3, or 4). Write the number of the correct answer in the brackets provided.

1. How many squares with side measurements of 5 centimeters are required to form a square whose perimeter is 2 meters?
    (1) 10   (2) 32
    (3) 40   (4) 100   (   )

2. A piece of wire, 8 feet long, is bent to form a square. What is the area enclosed by the square?
    (1) 4 square feet   (2) 16 square feet
    (3) 32 square feet   (4) 64 square feet   (   )

Study the figure and answer questions 3 and 4.

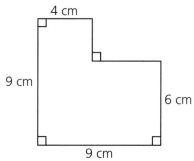

3. What is the area of the figure?
    (1) 90 cm²   (2) 66 cm²
    (3) 42 cm²   (4) 28 cm²   (   )

4. What is the perimeter of the figure?
    (1) 28 cm²   (2) 31 cm²
    (3) 36 cm²   (4) 90 cm²   (   )

The bar graph below shows the number of students in a given class who came late to school during a given week. Study it and answer questions 5 through 10.

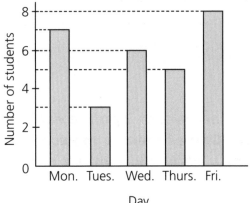

5. On which day did the largest number of students come late?
    (1) Friday   (2) Monday
    (3) Tuesday   (4) Wednesday   (   )

6. On which day did the smallest number of students come late?
    (1) Friday   (2) Monday
    (3) Tuesday   (4) Wednesday   (   )

7. How many students came late on Thursday?
    (1) 5   (2) 6
    (3) 7   (4) 8   (   )

8. How many more students came late on Monday than on Tuesday?
    (1) 3   (2) 4
    (3) 7   (4) 10   (   )

9. On which day were 7 students late?
    (1) Friday   (2) Monday
    (3) Tuesday   (4) Wednesday   (`   )

10. There were 40 students in the class. What fraction of the students in the class was late on Friday?
    (1) $\frac{1}{8}$ (2) $\frac{1}{5}$
    (3) $\frac{3}{10}$ (4) $\frac{4}{5}$ (   )

The bar graph below shows the number of Grade 5 students with perfect attendance for each of the four school terms. Study it and answer questions 11 through 15.

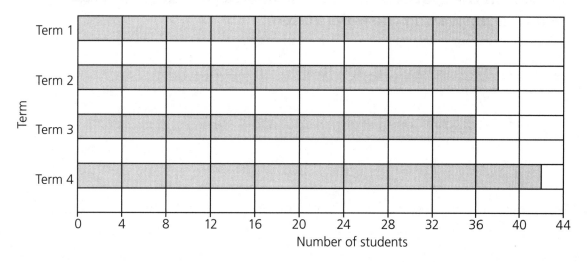

11. Which term has the best attendance?
    (1) Term 1   (2) Term 2
    (3) Term 3   (4) Term 4   (   )

12. How many students had perfect attendance in Term 1?
    (1) 36   (2) 37
    (3) 38   (4) 41   (   )

13. In which term did 36 students have perfect attendance?
    (1) Term 1   (2) Term 2
    (3) Term 3   (4) Term 4   (   )

14. There were 44 students in the class. How many did not have perfect attendance in Term 3?
    (1) 8   (2) 36
    (3) 37   (4) 41   (   )

15. How many more students had perfect attendance in Term 4 than in Term 1?
    (1) 1   (2) 4
    (3) 37   (4) 41   (   )

Math Practice the Singapore Way

Section B

Questions 16 through 30 are worth 2 points each. Write your answers in the spaces provided. For questions that require units, give your answers in the units stated.

16. The common factors of 15 and 18 are _____.

17. The first two common multiples of 2, 3, and 4 are _____ and

    _____.

18. What is the smallest number that can be divided by 3, 4, and 5? _____
    .

19. List the multiples of 9 between 50 and 90.

    _____

20. A rectangle has the same area as a square with a side measurement of 6 feet.

    The rectangle is 4 feet wide. Its length is _____ feet.

21. The length of a rectangle is twice its width. Its length is 10 feet.

    Find the perimeter of the rectangle. _____ feet

22. The perimeter of a rectangular piece of glass is 48 inches.

    Find its area if its length is 13 inches. _____ square inches

23. 20 handkerchiefs are made from a piece of cloth. Each side of a handkerchief is

    20 inches long. Find the area of the piece of cloth. _____ square inches

The graph below shows the number of visitors to the City Planetarium for a period of 5 days. Study it and answer questions 24 through 30.

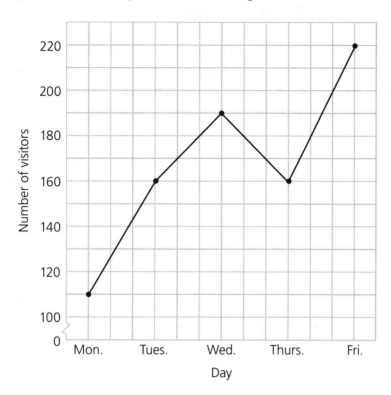

24. The smallest number of people visited the planetarium on _____.

25. The number of visitors was the same on _____ and _____.

26. There were _____ visitors on Wednesday.

27. There were _____ more visitors on Thursday than on Monday.

28. There were 220 visitors on _____.

29. There were 80 more visitors on _____ than on Monday.

30. The largest number of people visited the planetarium on _____.

Calculators are allowed in this section.

***For questions 31 through 41, show your work clearly in the space below each question, and write your answers in the spaces provided. The number of points you can earn is shown in brackets []
at the end of each question.***

The graph below shows the relation between the cost of a box of markers and the number of markers in the box. Use the graph to answer questions 31 and 32.

31. How many markers are in a $10-box?

    [2 points] _____

32. What is the cost of a box containing 16 markers?

    [2 points] _____

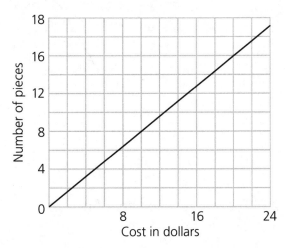

33. Δ Δ stands for 6 houses. Therefore, Δ Δ Δ stands for _____. [2 points]

One Country A dollar is approximately equal to 0.4 Country B dollars. The graph below shows their relationship. Use it to answer questions 34 and 35.

34. How many Country A dollars are equivalent to ten Country B dollars?

    [2 points] _____

35. Express A$15 in Country B dollars.

    [2 points] _____

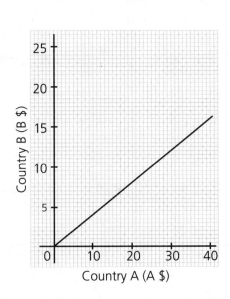

The graph below shows the amount of water that was collected from a leaking faucet over a period of time. Use it to answer questions 36 through 38.

36. How much water was collected in $2\frac{1}{2}$ hours? [2 points] _____

37. How long did it take to collect 100 milliliters of water?

    [2 points] _____

38. How long did it take to collect 700 milliliters of water?

    [2 points] _____

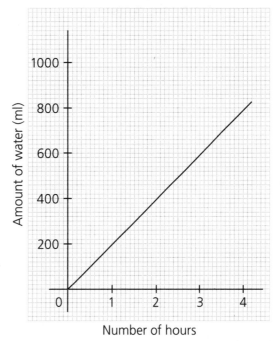

Number of hours

39. Pedro wants to tile his kitchen floor, which is 5 yards long and 3 yards wide. How much does he have to pay if tiling costs $24 per square yard? [4 points]

_____

40. A picture is 15 inches long and 10 inches wide. It is mounted on poster board measuring 17 inches by 12 inches. Find the area of the poster board that is not covered by the picture. [5 points]

_____

41. The area of a rectangular garden is 72 square yards. Its length is 9 yards. How much will it cost to fence it at $6 per yard? [5 points]

_____

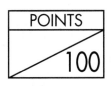

Section A

Questions 1 through 15 are worth 2 points each. For each question, four options are given. One of them is the correct answer. Make your choice (1, 2, 3, or 4). Write the number of the correct answer in the brackets provided.

1. Write 370,012 in words.
   (1) Thirty seven and twelve
   (2) Three thousand seven hundred and twelve
   (3) Thirty-seven thousand and twelve
   (4) Three hundred seventy thousand and twelve   ( )

2. Find the value of $6 + 16 \div 2 + 2$.
   (1) 16   (2) 12
   (3) 10   (4) 5   ( )

3. Write $\frac{2}{3}$ as a decimal, correct to 2 decimal places.
   (1) 0.6   (2) 0.7
   (3) 0.66   (4) 0.67   ( )

4. Multiply 4.3 by 100.
   (1) 0.043   (2) 43
   (3) 430   (4) 4,300   ( )

5. 3 tens, 5 hundredths is the same as _____.
   (1) 0.35   (2) 3.05
   (3) 30.05   (4) 30.5   ( )

6. 37% means _____.
   (1) $\frac{37}{10}$ (2) $\frac{37}{100}$
   (3) 37×100 (4) 37 out of 1 whole   ( )

7. Which one of the following is the greatest in value?
   (1) 8% of $1,000
   (2) 0.7 of $1,000
   (3) $\frac{3}{5}$ of $1,000
   (4) $\frac{6}{100}$ of $1,000   (   )

8. 2 kilograms, 20 grams written in kilograms is _____.
   (1) 2.002 kilograms
   (2) 2.02 kilograms
   (3) 2.2 kilograms
   (4) 20.2 kilograms   (   )

9. Which one of the following is equal to 60 grams?
   (1) $\frac{1}{20}$ kilograms
   (2) $\frac{1}{25}$ kilograms
   (3) $\frac{3}{50}$ kilograms
   (4) $\frac{3}{5}$ kilograms   (   )

10. Which one of the following is **incorrect**?
    (1) $\frac{21}{6} = 3\frac{2}{3}$
    (2) $6 \div 8 = \frac{3}{4}$
    (3) $\frac{4}{5} + \frac{1}{2} = 1\frac{3}{10}$
    (4) $1\frac{5}{8} - \frac{5}{6} = \frac{19}{24}$ (   )

11. Find the area of a square garden plot whose perimeter is 36 feet.
    (1) 6 square feet
    (2) 2 square feet
    (3) 24 square feet
    (4) 81 square feet   (   )

The bar graph below shows the grades, on a fifty-point scale, of four students who took a science test. Use it to answer questions 12 and 13.

12. Who earned the highest scores?
    (1) Andy
    (2) Tamara
    (3) Peggy
    (4) Paul   (   )

13. Who earned 35 points?
    (1) Andy
    (2) Tamara
    (3) Peggy
    (4) Paul   (   )

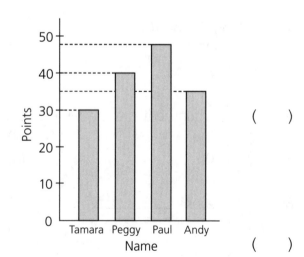

14. Joy's bowling score was 124 points. Lucia scored 6 points less than Joy. Chris scored half as many points as Joy. What was their total score?

(1) 130
(2) 192
(3) 242
(4) 304
( )

15. Linda bought $\frac{3}{4}$ pounds of flour. She used $\frac{1}{3}$ of it to bake a cake. She put the remainder equally in 4 bags. How many ounces of flour did each bag contain?

(1) 2 ounces
(2) 125 ounces
(3) 150 ounces
(4) 250 ounces
( )

Section B

Questions 16 through 30 are worth 2 points each. Write your answers in the spaces provided. For questions that require units, give your answers in the units stated.

16. Find the cost of parking a car for 2 hours, 20 minutes in a parking garage that charges $2 for the first hour and 80¢ for each subsequent $\frac{1}{2}$ hour or part of a half hour. $_____

17. Solve $\frac{5}{8} \times 4$. _____

18. A bowling team won 17 out of 20 games. What percentage of the games did it lose? _____%

19. Cheryl wants to share $\frac{3}{4}$ of a pound of melon seeds equally among 6 children. Each child will get _____ of a pound of melon seeds.

20. Mr. Wang's monthly income is $4,762. Round off his income for the first half of the year to the nearest $1,000. $_____

21. Fred has 270 rulers to tie in bundles. He ties 15 rulers in each bundle. How many bundles will he get? _____

22. Batiste owns a rectangular piece of land that is 170.2 feet long and 48.7 feet wide. How much shorter is the width than the length? _____ feet

23. $\angle x =$ _____°

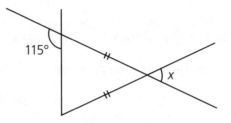

24. 5 pounds of beef costs $49. Find the cost of 3 pounds of beef. $_____

25. How many right angles does a rectangle have? _____

26. What number is 10,000 more than 218,643? _____

27. How many of these cubes will fill up a box that is 24 cm long, 18 cm wide, and 15 cm high? _____

28. There are 1,200 students in a school, and 28% of them wear glasses. Of those who do not wear glasses, $\frac{5}{8}$ are boys. How many boys do not wear glasses?

_____

29. The average weight of Dottie and Art is 61.4 pounds. Angel, who weighs 54.8 pounds, decides to join them. What will the average weight of the three people be? _____ pounds

30. Package A weighs 3 kilograms, 600 grams. Package B is 1.2 kilograms lighter than Package A. The ratio of the weight of Package A to the weight of Package B is _____.

Section C

For questions 31 through 38, show your work clearly in the space below each question, and write your answers in the spaces provided. The number of points you can earn is shown in brackets [] at the end of each question.

31. Jasmine packed 350 ounces of hazelnuts equally into each of 7 bags. She received an order for 274 bags from a customer. After making the delivery, she had 43,962 ounces of hazelnuts left. Find the weight of hazelnuts she had at first. [5 points]

_____

32. It costs $1,254 to fence a rectangular plot of land that is 18 yards long and 15 yards wide. How much does the fencing cost per yard? [5 points]

_____

33. One stick weighs 1 kilogram, 8 grams, and another stick weighs 475 grams. The weight of a third stick is 0.25 kilograms. What is their total weight? [5 points]

_____

34. A machine was used to pack mixed nuts into tins. Each filled tin weighs 273 ounces. Mr. Navararo received an order for 396 tins of mixed nuts. What was the total weight of mixed nuts needed? [5 points]

                    _____

35. The weights of Crates A, B, and C are in the ratio 6 : 5 : 3. The total weight of the three crates is 2,030 pounds. What is the total weight of Crate A and Crate C? [5 points]

                    _____

36. A square tile with a side measurement of 20 centimeters costs $2.30. How much will it cost to cover the floor of a room measuring 3 meters by 2 meters with square tiles? [5 points]

                    _____

37. Victor's monthly salary is $3,850. He saves 35% of it every month. How much money does he save in a year? [5 points]

                                                    _____

38. A factory packs 48 knives in a box. There are 1,100 boxes to fill. 46,455 knives have already been packed. How many more knives must be packed to fill the boxes? [5 points]

                                                    _____

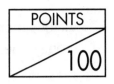
Section A

Questions 1 through 15 are worth 2 points each. For each question, four options are given. One of them is the correct answer. Make your choice (1, 2, 3, or 4). Write the number of the correct answer in the brackets provided.

1. 500,000 + 80,000 + 20 + 7 = _____
    (1) 5,827   (2) 58,027
    (3) 508,027   (4) 580,027   (   )

2. The value of the digit 5 in the ten thousands place is _____ times the value of the digit 5 in the tens place.
    (1) 10   (2) 100
    (3) 1,000   (4) 10,000   (   )

3. All the common factors of 12 and 18 are _____.
    (1) 1, 2, 3, and 6   (2) 2 and 3
    (3) 3 and 6   (4) 36 and 72   (   )

4. $\frac{3}{4}$ of one of the following figures is shaded. Which one is it?
    (1)   (2)

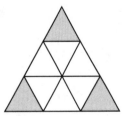
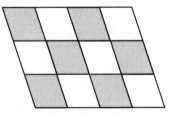

    (3)   (4)

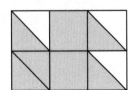
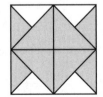

                                                                           (   )

5. There are _____ grams in $1\frac{1}{20}$ kilograms.
   (1) 1,050   (2) 150
   (3) 50   (4) $1\frac{1}{20}$ (   )

6. $\frac{39}{10}$ written as a decimal is _____.
   (1) 39.0   (2) 3.9
   (3) 0.39   (4) 0.039   (   )

7. Divide 0.3 by 10.
   (1) 0.003   (2) 0.03
   (3) 3   (4) 30   (   )

8. What number is one hundredth less than 3.406?
   (1) 2.406   (2) 3.309
   (3) 3.396   (4) 3.405   (   )

9. Six million and three thousand in numerals is _____.
   (1) 6,003   (2) 60,300
   (3) 600,300   (4) 6,003,000   (   )

10. Which one of the following has the smallest value?
    (1) $\frac{3}{4}$ (2) 0.6
    (3) 35%   (4) 50 out of 100   (   )

11. Half a pie is divided into 3 equal pieces. Each piece is _____ of the whole pie.
    (1) $\frac{1}{6}$ (2) $\frac{1}{5}$
    (3) $\frac{2}{3}$ (4) $\frac{3}{2}$ (   )

12. Which one of the following has the greatest volume?
    (1) A 4-inch cube
    (2) A box that is 3 inches long, 2 inches wide, and 4 inches high
    (3) A container that can hold 27 cubic inches of water
    (4) A box measuring 6 inches by 2 inches by 2 inches   (   )

Math Practice the Singapore Way

13. Marina's house is $\frac{7}{8}$ miles from school. She travels this distance twice daily.

What distance will she travel in five days?

(1) $1\frac{3}{4}$ miles   (2) $4\frac{3}{4}$ miles

(3) $8\frac{3}{4}$ miles   (4) $10\frac{7}{8}$ miles   ( )

14. The area of the shaded portion of the square is _____.
(1) 45.5 cm²
(2) 44.5 cm²
(3) 33 cm²
(4) 28 cm²

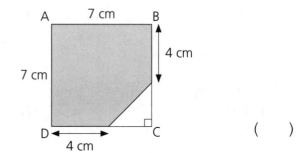

( )

15. ABC is a straight line. ∠c is equal to _____.
(1) 180° − ∠a − ∠b − ∠d
(2) ∠a + ∠b
(3) ∠a − ∠b
(4) ∠a − ∠d

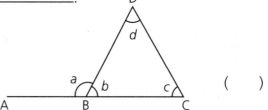

( )

Section B

Questions 16 through 30 are worth 2 points each. Write your answers in the spaces provided. For questions that require units, give your answers in the units stated.

16. Write 3.02 liters in milliliters. _____ milliliters

17. Divide 1,156 by 27. The answer is _____ with a remainder of _____

18. Find the product of 97 and 89 and give your answer, correct to the nearest hundred. _____

19. Write the sum of 7, 0.89, $\frac{3}{8}$, $1\frac{1}{4}$, and $\frac{9}{5}$ as a decimal. _____

20. Jackie jogged for $1\frac{1}{4}$ hours. He started at 11:50 a.m. At what time did he stop?

_____

21. Ms. Moore bought some fiction books and twice as many biography books for the school library. She bought 513 books altogether. How many biography books did she buy for the library? _____

22. Multiply 6.38 by 49. _____

23. Write 15 : 9 : 6 in the simplest form. _____

24. Solve $3\frac{5}{8} - 1\frac{5}{6}$. _____

25. Draw $\angle ABC = 55°$ and label it as $\angle b$.

26. PQR is a straight line.

    $\angle a = \frac{2}{5}$ of a right angle

    $\angle b =$ _____ of a right angle

    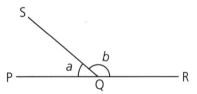

27. A typist can type 170 words in 5 minutes. How many words can she type in $\frac{1}{2}$ hour? _____

28. How many tiles, each measuring 10 centimeters by 6 centimeters, are needed to cover a floor measuring 4 meters by 3 meters? _____

The graph below shows the number of cars sold by a car salesperson for the first quarter of a given year. Use it to answer questions 29 and 30.

29. How many more cars were sold in February than in March?

_____

30. Find the total number of cars sold during the three months.

_____

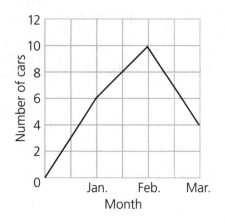

Section C

 Calculators are allowed in this section.

For questions 31 through 38, show your work clearly in the space below each question, and write your answers in the spaces provided. The number of points you can earn is shown in brackets [] at the end of each question.

31. A number, when divided by 25, gives a remainder of 23 and a quotient of 1,426. What is the number? [5 points]

_____

32. The rental charge for a car is $65.25 per day. Peter rented it for 2 weeks and was given a discount of $47. How much did he have to pay? [5 points]

_____

33. The perimeter of a rug is 14 yards, and its length is $4\frac{3}{4}$ yards. How much shorter is its width than its length? [5 points]

_____

34. Julio had $882 in his savings account. He added $65 to it every month for six months. He then drew out $\frac{1}{12}$ of his total savings to buy a pair of vases. What was the price of each vase? [5 points]

_____

35. On Saturday, Estrella spent $1\frac{3}{4}$ hours working in the garden, $\frac{5}{6}$ hours watching television, and $1\frac{2}{3}$ hours on her homework. How much time did she spend on these three activities? [5 points]

_____

36. Shaun saws off 80 centimeters from one end of a piece of brass rod that is 2.3 meters long. What is the length of the remaining piece of brass? Write your answer as a percent of 15 meters. [5 points]

_____

37. The average age of 12 children is 4.75 years. The average age of 8 of them is 5 years, 10 months. What is the average age of the remaining 4 children? [5 points]

_____

38. Ellen had 1,344 rubber bands. She put them in bags of 24 rubber bands each and sold them at $0.85 per bag. How much money would she receive if she sold 75% of the bags? [5 points]

_____

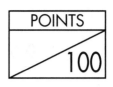

POINTS

/ 100

Section A

Questions 1 through 15 are worth 2 points each. For each question, four options are given. One of them is the correct answer. Make your choice (1, 2, 3, or 4). Write the number of the correct answer in the brackets provided.

1. Which one of the following is the smallest?
   (1) 389,700   (2) 408,700
   (3) 408,695   (4) 397,800   (   )

2. In the number 444,444, the value of the digit 4 in the thousands place is
   _____ times the value of the digit 4 in the ones place.
   (1) 10   (2) 100
   (3) 1,000   (4) 10,000   (   )

3. Which one of the following is **incorrect**?
   (1) $\frac{5}{2} = \frac{10}{4}$ (2) $\frac{6}{6} = 1$
   (3) $\frac{1}{4} = \frac{3}{8}$ (4) $\frac{7}{3} = 2\frac{1}{3}$ (   )

4. Which one of the following is **not** equal to $\frac{1}{4}$?
   (1) $3 \div 12$ (2) $\frac{5}{6} - \frac{7}{12}$
   (3) $\frac{7}{8} \times \frac{2}{7}$ (4) $\frac{3}{4} \div 6$ (   )

5. In 0.187, the value of the digit 8 is _____.
   (1) $\frac{8}{100}$ (2) $\frac{8}{10}$
   (3) 8   (4) 80   (   )

6. Percent means out of one _____.
   (1) hundredth   (2) hundred
   (3) tenth   (4) ten   (   )

7. 80% written as a decimal is _____.
   (1) 0.08   (2) 0.8
   (3) 8.0   (4) 80.0   (   )

8. Which one of the following is equivalent to 6 : 5?
   (1) 5 : 4   (2) 5 : 6
   (3) 12 : 9   (4) 18 : 15   (   )

9. How many times is the volume of a 10-foot cube as large as the volume of a 5-foot cube?
   (1) 2   (2) 4
   (3) 5   (4) 8   (   )

10. Bina divides $\frac{3}{4}$ of a block of butter into 6 equal pieces. What fraction of the whole block of butter will each piece be?
    (1) $\frac{1}{8}$ (2) $\frac{3}{4}$
    (3) $4\frac{1}{2}$ (4) $5\frac{1}{2}$ (   )

11. Which two lines are perpendicular to each other?
    (1) AE ⊥ DE
    (2) CB ⊥ AB
    (3) DC ⊥ BC
    (4) ED ⊥ CD   (   )

12. Sharon spent $\frac{1}{3}$ of her money in one store and $\frac{7}{15}$ of it in another. What fraction of her money did she have left?
    (1) $\frac{2}{15}$ (2) $\frac{1}{5}$
    (3) $\frac{1}{2}$ (4) $\frac{4}{5}$ (   )

13. Ms. Conner contributed $2, and each student in her class contributed 20¢ to a building fund. The class collected $9.80 altogether. How many students were in Ms. Conner's class?

 (1) 39   (2) 40

 (3) 49   (4) 59   (   )

Study the figure below and answer questions 14 and 15.

14. The area of the figure is _____.

 (1) 20 cm²
 (2) 36 cm²
 (3) 48 cm²
 (4) 96 cm²

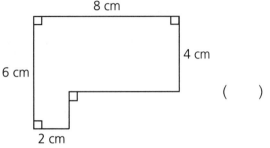

(   )

15. The perimeter of the figure is _____.

 (1) 20 cm   (2) 26 cm

 (3) 28 cm   (4) 50 cm   (   )

Section B

Questions 16 through 30 are worth 2 points each. Write your answers in the spaces provided. For questions that require units, give your answers in the units stated.

16. 2.45 kilograms is equal to _____ grams.

17. Solve 0.7 – 1.9 + 3.45, and give your answer as an improper fraction.

 _____

18. Round off 13.4317, correct to 2 decimal places. _____

19. $\frac{7}{8}$ written as a decimal is _____.

20. Round off 28,731, correct to the nearest ten. _____

Math Practice the Singapore Way

21. The sum of two numbers is 76. One of the numbers is 29. Find the product of the two numbers. _____

22. The volume of a rectangular box is 864 cubic inches. It is 9 inches long and 8 inches wide. Find its height. _____ inches

23. A machine can produce 4 buttons in 5 seconds. How many buttons can it produce in 1 minute? _____

24. One pencil costs 15¢. How many pencils can Miriam buy with 6 25¢ coins, 6 10¢ coins, and 6 5¢ coins?

_____

25. 11:35 p.m. is _____ minutes before midnight.

26. ABCD is a straight line.

    ∠p = _____ °

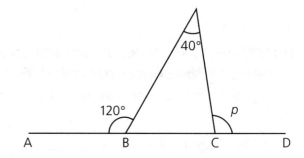

27. A tailor made 14 dresses. He used 3.5 meters of cloth for each dress. How much cloth did he use altogether? _____ kilometers

28. Dresses listed at $30 each are sold at a 20% discount. How much do 6 dresses cost? $_____

The bar graph below shows the number of stamps Kurt added to his collection for the first quarter of 2012. Use it to answer questions 29 and 30.

29. During which months did he collect the same number of stamps?

_____

30. How many stamps did he add to his collection in March?

_____

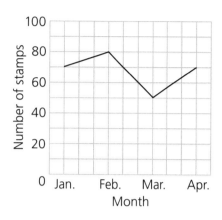

Section C

 Calculators are allowed in this section.

For questions 31 through 38, show your work clearly in the space below each question, and write your answers in the spaces provided. The number of points you can earn is shown in brackets [] at the end of each question.

31. Lucy bought three melons, which weighed $2\frac{1}{3}$ pounds, $1\frac{1}{2}$ pounds, and $1\frac{5}{6}$ pounds, respectively. What was the total weight of all three melons? [5 points]

_____

32. A car manufacturer ships 12,468 cars in a year. On average, how many cars does it ship every 3 months? [5 points]

_____

33. A ticket seller sold 129 $5 tickets and 210 $2 tickets. He received 12% commission on the sales. How much was his commission? [5 points]

_____

34. Eric drove $5\frac{3}{5}$ miles in the morning and $3\frac{7}{10}$ miles in the afternoon. How far short of 10 miles was the total distance he traveled? [5 points]

_____

35. A carton holds 24 cans of sardines. Each can costs $1.25. Find the cost of 4 cartons of sardines. [5 points]

_____

36. Sofia weighs $26\frac{1}{2}$ kilograms and Mary weighs $29\frac{3}{4}$ kilograms. Perla weighs 5 kilograms, 80 grams less than their average weight. What is Perla's weight? [5 points]

_____

37. Henry mixes 30 gallons of milk with 2 gallons of water. He then pours the mixture into 2-gallon bottles and sells them for $2.45 a bottle. How much money will he get? [5 points]

_____

38. Mr. Tomas had 12 gallons of orange juice. He served 45% of it to his guests at a party and used another 3.254 gallons to make fruit punch. How many gallons of orange juice were left? [5 points]

_____

UNITS

Test Yourself 5

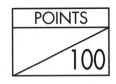

POINTS

/100

Section A

Questions 1 through 15 are worth 2 points each. For each question, four options are given. One of them is the correct answer. Make your choice (1, 2, 3, or 4). Write the number of the correct answer in the brackets provided.

1. In 105,876, which digit is in the ten thousands place?
    (1) 0   (2) 1
    (3) 5   (4) 8   (   )

2. A number, when rounded off to the nearest thousand, is 4,000. Which one of the following is the number?
    (1) 3,476   (2) 4,298
    (3) 4,710   (4) 5,627   (   )

3. 60 ÷ 1,000 = _____
    (1) 0.6   (2) 0.06
    (3) 0.006   (4) 0.0006   (   )

4. Which one of the following fractions has the largest value?
    (1) $\frac{5}{8}$ (2) $\frac{5}{6}$
    (3) $\frac{1}{2}$ (4) $\frac{2}{3}$ (   )

5. 4 ÷ 6 is **not** equal to _____.
    (1) $\frac{2}{3}$ (2) $\frac{1}{6}$ of 4
    (3) $\frac{1}{4}$ of 6   (4) $\frac{2}{6} + \frac{2}{6}$ (   )

6. What number is one tenth more than 10.296?
    (1) 10.297   (2) 10.306
    (3) 10.396   (4) 11.296   (   )

7. Which one of the following is **incorrect**?

   (1) $\frac{3}{4} + \frac{7}{12} = 1\frac{1}{3}$ (2) $3\frac{1}{2} - 1\frac{2}{3} = 1\frac{5}{6}$

   (3) $3\frac{1}{3} - 1\frac{7}{8} = 1\frac{11}{24}$ (4) $\frac{2}{3} \div 2 = 1\frac{1}{3}$ (   )

The bar graph below shows the number of Grade 5 students in a certain school who took part in an art competition. Use it to answer questions 8 and 9.

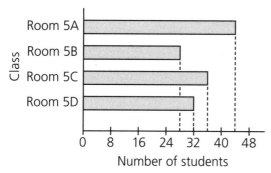

8. How many more students from Room 5C than Room 5D took part in the art competition?

   (1) 4   (2) 8

   (3) 28   (4) 52   (   )

9. Which class had the most pupils taking part in the competition?

   (1) Room 5A   (2) Room 5B

   (3) Room 5C   (4) Room 5D   (   )

10. 8% of the distance between two towns is 4 miles. How far apart are they?

   (1) 12 miles   (2) 32 miles

   (3) 50 miles   (4) 368 miles   (   )

Study the figure below, which is made up of rectangles. Use it to answer questions 11 and 12.

11. The perimeter of the figure is _____.

   (1) 27 cm

   (2) 32 cm

   (3) 33 cm

   (4) 36 cm

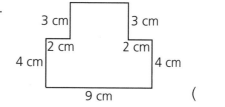

(   )

Math Practice the Singapore Way

12. The area of the figure is _____.
    (1) 42 cm²   (2) 48 cm²
    (3) 51 cm²   (4) 84 cm²   (   )

13. A carpet measuring 15 feet by 12 feet is placed on the floor of a room, leaving
    a border 2 feet wide all around it. The area of the border is _____.
    (1) 58 square feet   (2) 63 square feet
    (3) 92 square feet   (4) 124 square feet   (   )

14. A family of three earns an average of $654 a month. Find their total income for
    one year.
    (1) $1,962   (2) $2,616
    (3) $7,848   (4) $23,544   (   )

15. I spent 30% of my money in one store and 0.5% of it in another store.
    What fraction of my money did I have left?
    (1) $\frac{3}{20}$ (2) $\frac{1}{5}$
    (3) $\frac{7}{20}$ (4) $\frac{4}{5}$ (   )

Section B

**Questions 16 through 30 are worth 2 points each. Write your
answers in the spaces provided. For questions that require units,
give your answers in the units stated.**

16. The best approximation for 39 × 4.01 is _____.

17. Write 45 minutes as a fraction of 2 hours. _____

18. 0.8% of a number is 16. What is the number? _____

19. Write 6 meters in kilometers. _____ kilometers

20. Find the product of 6.2 and 30. _____

21. The volume of a cube is 343 cubic centimeters. Its edge is _____ centimeters.

22. Complete the following:
70,000, 68,000, 66,000, 64,000, _____, _____

23. The area of the shaded portion of the rectangle is _____ cm².

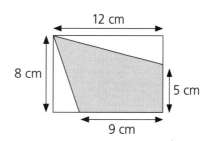

24. Find the value of 18 − 3 × 6 + 4. _____

25. Mr. Lewis weighs 153 pounds. Mr. Lewis's son weighs $\frac{1}{3}$ as much as Mr. Lewis. How much heavier is Mr. Lewis than his son? _____ pounds

26. Each side of a triangle is 5 centimeters. Each angle of the triangle is _____°.

27. Dylan's family uses 500 milliliters of milk per day. How many liters of milk will Dylan's family use in the month of December? _____ liters

28. Mr. Hammond shares 390 pencils among 8 boys and 13 girls so that each boy gets 2 pencils more than each girl. How many pencils will be left over? _____

29. A water tank is 4 feet long, 3 feet wide, and 2 feet high. It is $\frac{1}{8}$-filled with water. How much more water can it hold? _____ cubic feet

30. Draw a triangle ABC in which AB = 6 centimeters, BC = 4 centimeters, and ∠ABC = 35°. Measure ∠CAB.

∠CAB = _____°

Section C

 Calculators are allowed in this section.

For questions 31 through 38, show your work clearly in the space below each question, and write your answers in the spaces provided. The number of points you can earn is shown in brackets []
at the end of each question.

31. Sue bicycled $2\frac{3}{4}$ miles to her aunt's house. She then bicycled $1\frac{5}{8}$ miles to school before heading to the mall, which was $2\frac{1}{2}$ miles away. What was the total distance that she bicycled? [5 points]

_____

32. Ms. Young bought 25 pieces of garden hose. Each piece was 0.28 yards less than 2 yards. How many yards of garden hose did she buy in all? [5 yards]

_____

33. Pencils are sold in bundles of 12 for $2.75. Omar needs 250 pencils for his office. How much will the pencils cost? [5 points]

_____

34. A tank can hold 20 quarts of water. There are $2\frac{5}{8}$ quarts of water in it. Aaron pours $12\frac{3}{4}$ quarts of water into it. How much more water does he need to pour in to fill it up? [5 points]

_____

35. Ms. Nelson bought 11 boxes of wheat flour, each weighing 0.75 pounds. She used $4\frac{3}{8}$ pounds of it to fry some prawns. How much wheat flour was left? [5 points]

_____

36. Amy, Kelly, and Sandi share 2,466 marbles in the ratio 3 : 7 : 8. How many marbles should Kelly give Amy so that they have the same number of marbles? [5 points]

_____

37. Desiree took $1\frac{3}{5}$ hours to finish her math homework. She took another $\frac{2}{3}$ hours to complete a journal entry before eating dinner at 7:25 p.m. At what time did she start working on her math homework? [5 points]

_____

38. Nicholas bought a sofa worth $1,350. He made a down payment of $290 and arranged to pay the balance in monthly installments of $53 per month. How many years will it take him to complete the full payment? [5 points]

_____

Section A

Questions 1 through 15 are worth 2 points each. For each question, four options are given. One of them is the correct answer. Make your choice (1, 2, 3, or 4). Write the number of the correct answer in the brackets provided.

1. The value of the digit 9 in 897,600 is _____.
    (1) 9 hundreds   (2) 9 thousands
    (3) 90 thousands   (4) 900 thousands   (   )

2. 10,000,000 is _____ times greater than 10,000.
    (1) 100   (2) 1,000
    (3) 10,000   (4) 100,000   (   )

3. Multiply 1,428 by 68 and give your answer, correct to the nearest thousand.
    (1) 19,992   (2) 20,000
    (3) 97,000   (4) 98,000   (   )

4. Which one of the following is **incorrect**?
    (1) 1.5 kg = 150 g   (2) 30 cm = 0.3 m
    (3) 2 L, 500 ml = 2.5 L (4) 1.8 m = 180 cm   (   )

5. From a coil of rope, Jay used $1\frac{1}{4}$ meters and cut the remainder into 10 pieces, each 0.35 meters long. What was the length of the original coil?
    (1) 1.6 meters   (2) 2.25 meters
    (3) 3.5 meters   (4) 4.75 meters   (   )

6. Write 0.3 as a percentage.
    (1) 0.03%   (2) 0.3%
    (3) 3%   (4) 30%   (   )

Math Practice the Singapore Way   **125**

7. Express 40% as a fraction in its simplest form.

   (1) $\frac{1}{4}$ (2) $\frac{4}{10}$

   (3) $\frac{2}{5}$ (4) $\frac{4}{100}$ ( )

8. $\frac{1}{3}$ of one of the following figures is shaded. Which one is it?

   (1) (2)

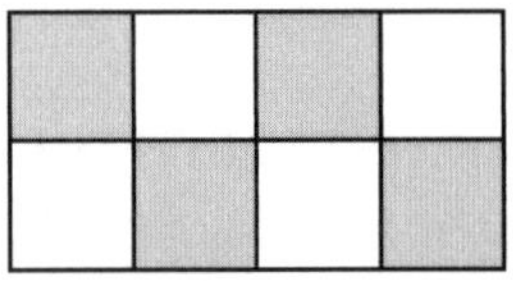

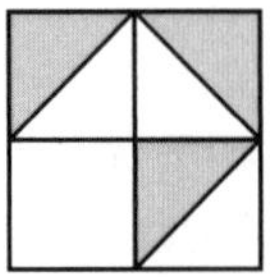

   (3) 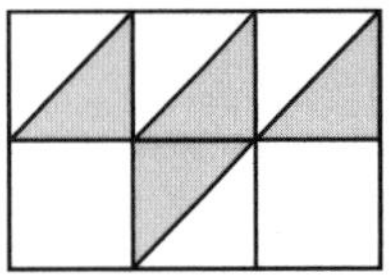 (4) ( )

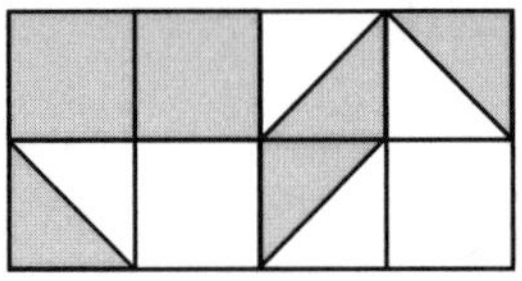

9. Alonso spent $\frac{1}{4}$ of his money on a watch and gave $\frac{1}{12}$ of it to his grandmother.

   He then had $120 left. What was the cost of the watch?
   (1) $15 (2) $45
   (3) $180 (4) $240 ( )

10. Which decimal reflects the shaded portion of the figure? 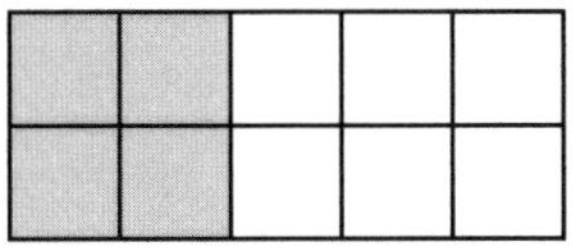
    (1) 0.04
    (2) 0.4
    (3) 4.0
    (4) 4 ( )

11. A square has the same area as a triangle whose base is 8 inches and height is 25 inches. Each side of the square is _____ inches long.
    (1) 10 (2) 25
    (3) 50 (4) 100 ( )

12. A piece of wallpaper was 3.75 yards long and 0.2 yards wide. Tom cut and pasted 1.75 yards of it on his shelf. What was the area of the piece of wallpaper that was not used?
    (1) 40 square yards (2) 400 square yards
    (3) 4,000 square yards (4) 40,000 square yards ( )

13. A rectangular wooden block measuring 20 inches by 16 inches by 12 inches is cut into cubes with side edge measurements of 2 inches. How many cubes will there be?

    (1) 480   (2) 960

    (3) 1,920   (4) 3,840   (   )

The graph below shows the height of four shelves. Study it and answer questions 14 and 15.

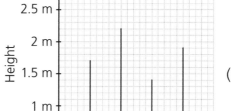

14. What is the height of Shelf A?

    (1) 1.4 meters

    (2) 1.7 meters

    (3) 1.8 meters

    (4) 2.2 meters   (   )

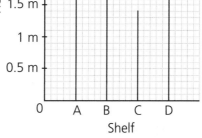

15. Which shelf has a height of 1.9 meters?

    (1) A

    (2) B

    (3) C

    (4) D   (   )

Section B

Questions 16 through 30 are worth 2 points each. Write your answers in the spaces provided. For questions that require units, give your answers in the units stated.

16. The whole number that comes just before 378,000 is _____.

17. 286,746 people turned up to watch a parade. Round off this number to the nearest thousand. _____

18. One hundredth less than 0.2 is _____.

19. $\frac{2}{3}$ of $\frac{9}{10}$ expressed as a decimal is _____.

20. 2 hours, 40 minutes is $\frac{5}{8}$ of _____ hours.

21. 4.3 meters written in centimeters is _____ centimeters.

22. What fraction of both circles is shaded? Write your answer as an improper fraction.

    _____

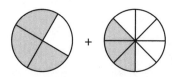

23. Find ∠n. _____°

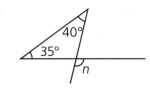

24. APB is a straight line.

    ∠QPB = _____°

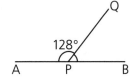

25. 3 : 4 = 9 : ☐. The missing number in the box is _____.

26. 5 eggs cost $0.70. How much will Rachel have to pay for 40 eggs?

    $_____

27. Debra bought 3 packages of hot dogs for $6.72. There were 8 hot dogs in each package. What was the cost of each hot dog? _____¢

28. The average income of a family of three is $1,430. Two of them earn $1,600 and $870. Find the income of the third person. $_____

29. It took Kenneth 55 minutes to paint a gate and $\frac{1}{4}$ hour to paint a stool. He finished the two jobs at 12:30 p.m. At what time did he begin to paint?

    _____

30. There were 16 rows of plants in a garden. There were 32 plants in each row. $\frac{11}{16}$ of them were flowering plants. How many of them were nonflowering plants? _____

Section C

For questions 31 through 38, show your work clearly in the space below each question, and write your answers in the spaces provided. The number of points you can earn is shown in brackets [] at the end of each question.

31. Poles A, B, and C were placed end to end in a straight line. Pole A is $1\frac{1}{6}$ feet long, Pole B is $4\frac{5}{6}$ feet long, and Pole C is $2\frac{5}{8}$ feet long. What is the length of the three poles altogether? [5 points]

_____

32. Caleb mixed 4.24 kilograms of potatoes, 1.57 liters of water, and 80 grams of curry powder with 8.42 kilograms of meat to cook curry for 45 people. How much curry, on average, did he estimate that each person would eat? [5 points]

_____

Math Practice the Singapore Way

33. Mr. Basco had $180. He paid $23.50 for a sweater, $18.50 for a raincoat, and spent 25% on some books. He used the remaining amount of money to buy some paper for 15 students. How much, on average, did he spend on each student? [5 points]

_____

34. Andrew bought $1\frac{1}{2}$ gallon of paint. He used $\frac{3}{4}$ gallon to paint his fence and $\frac{1}{3}$ of the remainder to paint a bench. How many gallons of paint did he have left? [5 points]

_____

35. A rectangular fish tank is 40 centimeters long and 30 centimeters wide. It contains water that is 8 centimeters in height. When a piece of metal is dropped into the water, the height of the water level becomes 20 centimeters. What is the volume of the piece of metal? [5 points]

_____

36. Zoe deposits $50,000 in a bank. The bank pays 7% interest per year. She withdraws the interest every year. How much interest will she withdraw after 8 years? [5 points]

_____

37. The ratio of the number of craft books to the number of art books to the number of music books in a bookstore is 3 : 7 : 4. There are 1,428 books altogether. How many more art books than music books are there? [5 points]

_____

38. For a field trip, a student pays 80¢ for bus fare, $1.20 for food and drinks, and 60¢ for admission to an aquarium. How much will the field trip cost if there are total of 45 children? [5 points]

_____

Section A

Questions 1 through 15 are worth 2 points each. For each question, four options are given. One of them is the correct answer. Make your choice (1, 2, 3, or 4). Write the number of the correct answer in the brackets provided.

1. The monthly expenditure of a factory is $299,750. Round off this amount to the nearest $1,000.
    (1) $29,000   (2) $299,000
    (3) $300,000   (4) $399,000   (   )

2. Which one of the following pairs of angles can be found in an isosceles triangle?
    (1) 60°, 60°   (2) 55°, 70°
    (3) 30°, 60°   (4) 20°, 90°   (   )

3. Simplify $15 + 30 - 12 \div 3$.
    (1) 14   (2) 39   (3) 41   (4) 44   (   )

4. $\frac{47}{1000} + \frac{19}{100} + \frac{3}{10} = $ _____
    (1) 0.537   (2) 0.69
    (3) 0.96   (4) 4.893   (   )

5. Which one of the following is **incorrect**?
    (1) $1\frac{1}{4}$ hours = 75 minutes   (2) $1\frac{1}{3}$ of a right angle = 130°
    (3) $1\frac{1}{2}$ kilograms = 1,500 grams   (4) $2\frac{1}{2}$ meters = 250 centimeters   (   )

6. Which one of the following is closest to the product of 53 and 7?
   (1) 350   (2) 370
   (3) 420   (4) 500   (   )

7. The difference between $3\frac{5}{12}$ and $5\frac{3}{4}$ is _____.
   (1) $2\frac{1}{4}$ (2) $2\frac{1}{3}$
   (3) $8\frac{8}{16}$ (4) $9\frac{1}{6}$ (   )

8. 0.32 written as a fraction in its lowest terms is _____.
   (1) $\frac{32}{10}$ (2) $\frac{32}{100}$
   (3) $\frac{8}{25}$ (4) $\frac{4}{5}$ (   )

9. What is the value of the digit 5 in 3.857?
   (1) 5 hundredths   (2) 5 ones
   (3) 5 tens   (4) 5 tenths   (   )

10. The area of the figure is _____.
    (1) 18 cm²
    (2) 33 cm²
    (3) 42 cm²
    (4) 48 cm²   (   )

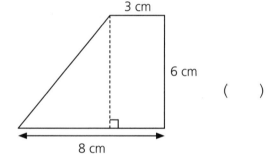

3 cm

6 cm

8 cm

11. Amy had $2\frac{1}{2}$ yards of lace. She cut it into 10 equal pieces. What is the length of each piece of lace?

    (1) 0.25 yards   (2) 7.5 yards

    (3) 12.5 yards   (4) 25 yards   (   )

12. Janie's father is 50 years old. He is 8 years older than 3 times Janie's age. How old is Janie?
    (1) 14 years old   (2) 24 years old
    (3) 26 years old   (4) 42 years old   (   )

13. Margie sold 168 lottery tickets. She had 240 tickets at first. What percentage of her tickets did she sell?
    (1) 0.7%   (2) 1.68%
    (3) 30%   (4) 70%   (   )

The graph below shows the number of $5, $10, $15, and $25 tickets sold at a circus. Use the graph to answer questions 14 and 15.

14. What is the total number of tickets sold?
    (1) 1,000
    (2) 2,500
    (3) 8,800
    (4) 27,000   (   )

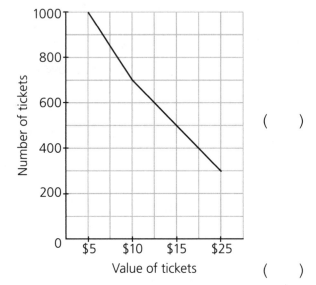

15. How much money was collected from the sale of the $25 tickets?
    (1) $75
    (2) $300
    (3) $5,000
    (4) $7,500   (   )

Section B

Questions 16 through 30 are worth 2 points each. Write your answers in the spaces provided. For questions that require units, give your answers in the units stated.

16. Write 1,000,000 in words. _____

17. Write the number between 40 and 60 of which both 4 and 7 are factors.

    _____

18. Solve $\frac{9}{10} \times \frac{5}{6}$. _____

19. Solve $4\frac{1}{9} - 1\frac{5}{6}$. _____

20. ABC is a straight line.

    ∠p = _____°

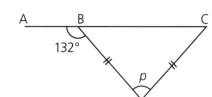

21. There were 400 apples in a basket. 10% were bad. $\frac{1}{20}$ of the remainder were

    eaten by rats. How many good apples were left? _____

22. 0.7% of a sum of money is $68.95. What is 20% of the sum of money?

    $_____

23. The area of a square is 36 square inches. Its perimeter is _____ inches.

24. Joan's mother is 150 centimeters tall. Joan's height is $\frac{2}{3}$ of her mother's height.

    How tall will Joan be if she grows another 2 centimeters? _____ meters

25. The average age of two boys is 7 years, 6 months. A third boy aged 7 years,
    9 months joins them. What is the average age of the three boys?

    _____ years _____, months

26. A tractor tank contains 80 liters of diesel fuel. It uses 15 liters of diesel fuel per
    hour. How much more fuel must be poured into the tank so that the tractor

    can operate for 10 hours? _____ liters

27. A rectangular tank, measuring 30 centimeters by 10 centimeters, contains
    7.2 liters of water. Find the height of the water level in the tank.

    (1 liter = 1,000 cubic centimeters) _____ centimeters

28. The total thickness of a stack of 5 pieces of cardboard is 1 centimeter.
    What is the total height of a stack of 42 pieces of cardboard?

    _____ centimeters

29. Joseph wants to buy 8 5¢ stamps and 15 10¢ stamps. How much change will

    he get if he gives the postal clerk a $20 bill? $_____

Math Practice the Singapore Way

30. A box with 8 books weighs 7.15 kilograms. The books are of equal weight and the weight of the empty box is 0.91 kilograms. What is the weight of each book? _____ kilograms

Section C

 Calculators are allowed in this section.

For questions 31 through 38, show your work clearly in the space below each question, and write your answers in the spaces provided. The number of points you can earn is shown in brackets [] at the end of each question.

31. A tank containing 44.82 liters of water can be emptied in 18 minutes. How much water can be emptied from the tank in 5 minutes? [5 points]

_____

32. The ratio of the number of chickens to the number of ducks on a farm was 15 : 2. After 1,078 chickens were sold, the ratio became 2 : 1. How many ducks were there at first? [5 points]

_____

33. Cherie bought 18 rolls of wallpaper at $24.50 a roll. She also bought 3 boxes of toothpaste at $2.30 a box. She had $52.10 left. How much did she have at first? [5 points]

_____

34. Alex bought a used refrigerator for $1,725. He spent 15% of that amount on repairs. How much did the refrigerator cost him altogether? [5 points]

_____

35. A carpenter bought $3\frac{1}{2}$ kilograms of nails. He used $1\frac{5}{8}$ kilograms of them. Each nail weighed 15 grams. How many nails did he have left? [5 points]

_____

36. The area of a rectangle 35 centimeters long is 2,450 square centimeters. What is the perimeter of the rectangle? [5 points]

_____

37. The ratio of the number of women to the number of men at a concert was 5 : 13. There were 1,092 men. How many women were there at the concert? [5 points]

_____

38. A window requires 3.8 meters of curtain material while a bathroom shower requires 1.9 meters. How much curtain material will be needed for 18 windows and 5 bathroom showers? [5 points]

_____

Section A

Questions 1 through 16 are worth 2 points each. For each question, four options are given. One of them is the correct answer. Make your choice (1, 2, 3, or 4). Write the number of the correct answer in the brackets provided.

1. The factors of 10 are _____.
   (1) 2   (2) 2 and 5
   (3) 1, 2, 5, and 10   (4) 10, 20, and 30   (   )

2. How much more than 504,876 is 505,876?
   (1) 10   (2) 100
   (3) 1,000   (4) 10,000   (   )

3. How many 16s are in 96?
   (1) 1   (2) 6
   (3) 80   (4) 1,536   (   )

4. A 4-sided figure with one pair of parallel sides is called a _____.
   (1) parallelogram   (2) rectangle
   (3) rhombus   (4) trapezoid   (   )

5. You are facing northwest. If you turn 90° counterclockwise, you will face
   _____.
   (1) northeast   (2) northwest
   (3) southeast   (4) southwest   (   )

6. Solve $\frac{3}{4} \times \frac{5}{6}$.
   (1) $\frac{5}{8}$ (2) $\frac{8}{10}$
   (3) $\frac{9}{10}$ (4) $\frac{15}{24}$ (   )

7. Which one of the following fractions is the largest?

    (1) $\frac{3}{8}$ (2) $\frac{1}{2}$

    (3) $\frac{4}{5}$ (4) $\frac{7}{10}$ (   )

8. Write $\frac{3}{8}$ as a decimal.

    (1) 0.375   (2) 0.3
    (3) 3.75   (4) 8.3   (   )

10. The closest approximate value of 27.8 × 16.1 is _____.
    (1) 27 × 16   (2) 27 × 17
    (3) 28 × 16   (4) 28 × 17   (   )

11. Which one of the following is **not** equal to 4%?
    (1) 0.4   (2) 4 out of 100

    (3) $\frac{1}{25}$ (4) $\frac{4}{100}$

12. How much will it cost to bind a 4-meter-long and 2-meter-wide rug with a ribbon that costs $3.80 a meter?
    (1) $15.20   (2) $22.80
    (3) $30.40   (4) $45.60

13. The area of the triangle is _____.
    (1) 32 cm²
    (2) 126 cm²
    (3) 252 cm²
    (4) 504 cm²   (   )

    14 cm | 18 cm

14. How many cubes with a side measurement of 2 centimeters are equivalent in volume to a rectangular box measuring 4 centimeters by 6 centimeters by 2 centimeters?
    (1) 6   (2) 12
    (3) 24   (4) 48   (   )

The figures in question 14 are not drawn to scale.

15. Which one of the following is an isosceles triangle?

(1)
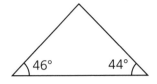

(2)
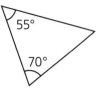

(3)
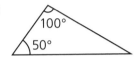

(4)
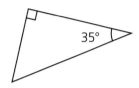

(   )

16. At lunch, Ivy put 28 sandwiches on the table. After lunch, there were 4 sandwiches left. 8 people were at the table, and they ate an equal number of sandwiches. How many did each person eat?

(1) 3

(2) $3\frac{1}{2}$

(3) 7

(4) 16

(   )

Section B

Questions 16 through 31 are worth 2 points each. Write your answers in the spaces provided. For questions that require units, give your answers in the units stated.

17. Evaluate 34 × 2 − 16 × 4 + 7. _____

18. What number multiplied by 38 will give 6,422? _____

19. Find the sum of $\frac{3}{4}$, $\frac{7}{8}$, and $\frac{5}{12}$. _____

20. Arrange the following decimals in order, beginning with the largest:
0.88, 0.8, 0.08, 0.808

_____ , _____ , _____ , _____

21. The first four multiples of 5 are _____ .

22. Sandra knits a scarf that measures 30 centimeters by 28 centimeters from 11:45 a.m. to 2:45 p.m. How many square centimeters does she knit in 1 hour? _____ square centimeters

23. One pencil costs $0.20, and one ruler costs $0.30. Jane spent $3.20. She bought 7 pencils and _____ rulers.

24. It takes $\frac{3}{4}$ hours to roast a chicken. Gwen has already roasted the chicken for 20 minutes. It will take _____ more minutes for the chicken to be fully roasted.

25. Martha has 8 identical vases that weigh $3\frac{2}{5}$ kilograms altogether. What is the weight of each vase? _____ grams

26. A square carpet with a side measurement of 6 meters is laid on the floor of a room measuring 15 meters by 12 meters. What percentage of the area of the floor is not covered by the carpet? _____%

27. David poured 9.7 quarts of oil into 4 bottles of the same size and had 3.9 quarts left. How much oil was did he pour into each bottle? _____ quarts

28. 4 meters of lace are required to go around a rectangular tablecloth that is 80 centimeters wide. How long is the tablecloth? _____ centimeters

29. There are 18 blue pencils and 24 red pencils in a box. Write the ratio of the number of blue pencils to the number of red pencils in the simplest form.

_____

The bar graph shows the amount of money collected at a parking garage in one week. Use it to answer questions 29 and 30.

30. What was the largest amount

    collected? $_____

31. On which day was $250 collected?

    _____

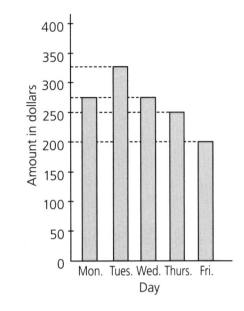

Section C

 Calculators are allowed in this section.

For questions 32 through 39, show your work clearly in the space below each question, and write your answers in the spaces provided. The number of points you can earn is shown in brackets [] at the end of each question or part of a question.

32. After James spent $6.45 at a farmer's market and $\frac{1}{12}$ of his remaining amount of money at another farmer's market, he had $19.25 left. How much did he have at first? [5 points]

    _____

33. Ms. Olsen and her son inherited $85,400. After taxes amounting to 19% of the sum of money had been deducted, Ms. Olsen received 65% of that amount and her son received the rest. How much did her son receive? [5 points]

_____

34. A carpet costs $5.99 per square meter. It is placed on the floor of a room measuring 7 meters by 5 meters. A $\frac{1}{2}$-meter-wide margin is left all around it. How much will it cost? [5 points]

_____

35. Cathy bought 12 bath towels at $3.50 each and 15 face towels at $0.95 each. The total amount spent was $\frac{5}{8}$ of the amount of money she had. How much did she have at first? [5 points]

_____

© 2012 Marshall Cavendish Corporation

36. The time limit on a math test was $1\frac{3}{4}$ hours. Jennie completed it 15 minutes before time was up.
    (a) How long did she take to complete the test? [2 points]
    (b) What fraction of the time did she take to complete the test? [3 points]

(a) _____

(b) _____

37. A box and its contents weigh $19\frac{3}{5}$ kilograms. The empty box weighs $1\frac{17}{20}$ kilograms. What is the weight of the contents in 6 identical boxes?
    [5 points]

Math Practice the Singapore Way

38. A bag weighs 2.842 pounds. A second bag is twice as heavy as the first one. A third bag is 0.245 pounds lighter than the second bag. What is the average weight of the three bags? [5 points]

_____

39. A tank is 28 centimeters long and 15 centimeters wide. It is filled with water to a height of 6 centimeters. Felicia puts a large rock into it and the height of the water level rises to 20 centimeters. What is the volume of the rock? [5 points]

_____

Answers

Unit 1 NUMBERS TO TEN MILLION

1. (1) 2. (1) 3. (4) 4. (3)

5. (2) 6. (4) 7. (3) 8. (4)

9. (3) 10. (3) 11. (2) 12. (3)

13. (4) 14. (2)

15. 15,080   16. 80,000

17. 27,800 − 26,800 = 1,000
    28,800 + 1,000 = 29,800
    29,800 + 1000 = 30,800

18. 48,934   19. 19,900

20. 100

21. 31,548, 2,888, 2,874, 900

22. Four million, one hundred, sixty thousand, and twelve

23. 34,875

24. 4,290, 2,940, 2,490, 2,049

25. 6,012,000

Unit 2 APPROXIMATION AND ESTIMATION

1. (3) 2. (4) 3. (3) 4. (2)

5. (4)

6. 5 zeros   7. 6,000

8. 38,000   9. 184,300

10. 107,000

11. 1,712 ≈ 2,000   2,000
    5,135 ≈ 5,000   5,000
    3,264 ≈ 3,000 + 3,000
                     10,000

Unit 3 THE FOUR OPERATIONS OF WHOLE NUMBERS

1. (3) 2. (4) 3. (1) 4. (2)

5. (1) 6. (3) 7. (2) 8. (3)

9. (4) 10. (3) 11. (4)

12. 58,682   13. 72

14. 98   15. 88

16. 50   17. 26

18. 3,976   3,976 ≈ 4,000   4,000
    × 9   × 9
    35,784   36,000

19. 15,963   15,963 ≈ 16,000   16,000
    −9,374   9,374 ≈ 9,000   −9,000
    6,589   7,000

20. 8   21. 3.85

22. 2 hours, 50 minutes − 1 hour
    = 1 hour, 50 minutes ≈ 2 hours
    $3 + $2.50 × 2 = $8

23. 700   24. 3,000

25. 700 − 250 = 450 ounces ≈ 500 ounces
    $17 + $2 × 2 = $21

26.

    2 units → 30
    1 units → 30 ÷ 2 = 15
    7 units → 15 × 7 = 105
    105 ÷ 3 − 30 = 5
    He must give 5 marbles to Ben.
    105 ÷ 3 − 15 = 20
    He must give 20 marbles to Bob.

27. 18 × 24 − 38 = 394
    394 cans were used during the party.

28. 1,330 ÷ 38 = 35
    Fran can make 35 necklaces.
    1026 ÷ 38 = 27
    Her sister can make 27 necklaces.
    35 − 27 = 8
    Fran can make 8 more necklaces than her sister.

29. $2,200 − $1,180 = $1,020
    $1,020 ÷ 12 = $85
    He had to pay $85 each month.

30. 1 hour = 60 minutes
    4,800 ÷ 60 = 80
    It can print 80 pages per minute.

31. 900 − 250 = 650 grams ≈ 750 grams
$20 + $3.50 × 3 = $30.50
He paid $30.50.

32. $18 − $5.40 = $12.60 = 1,260 cents
1,260 ÷ 90 = 14
She needs to save for another 14 weeks.

33. 1,120 ÷ 80 = 14
It makes 14 stitches per centimeter.
1.3 meter = 130 centimeters
14 × 130 = 1,820
It will make 1,820 stitches.

Units 1–3   ASSESS YOURSELF 1

1. (1)   2. (3)   3. (3)   4. (4)

5. (2)   6. (4)   7. (1)   8. (4)

9. 100,000   10. 73,200

11. 10,000   12. 150

13. One hundred ninety thousand and twelve

14. 8

15. Total number of T-shirts
= 12 × 4 = 48
$715.20 ÷ 48 = $14.90

16. 396 − 15 − 14 = 367
There were 367 students.

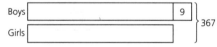

2 u → 367 − 9 = 358
1 u → 358 ÷ 2 = 179
There were 179 girls.

17.

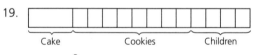

4 u → 78 − 6 = 72 pounds
1 u → 72 ÷ 4 = 18 pounds
2 u → 18 × 2 = 36 pounds
The weight of Bag A is 36 kilograms.
18 + 6 = 24 kilograms
The weight of Bag C is 24 kilograms.
36 − 24 = 12 kilograms
The difference in weight between Bag A and
Bag C is 12 kilograms.

18. $12,236 − $3,500 = $8,736
12 × 2 = 24 months
$8,736 ÷ 24 = $364
He paid $364 each month.

19. Total no. of handbags in 4 containers
= 1,428 × 4 = 5,712
Remaining no. of containers = 8 − 4 − 2 = 2
Total no. of handbags in last 2 containers
= (1,428 − 243) × 2 = 2,370
5,712 + 795 + 2,370 = 8,877
8,877 handbags were shipped altogether.

Unit 4   FRACTIONS

1. (3)   2. (3)   3. (2)   4. (4)

5. (4)   6. (4)   7. (1)   8. (2)

9. $\frac{1}{6}$ 10. 20

11. 7

12. $6\frac{1}{4} - 1\frac{1}{2} + 1\frac{2}{5} = 6\frac{3}{20}$ kg = 6 kg 150 g

13. $4\frac{4}{15}$ 14. $\frac{5}{8}, \frac{2}{3}, \frac{3}{4}, \frac{5}{6}$

15. $\frac{4}{15}$

16.

Food		Transport		$23

1 u → $23
6 u → $23 × 6 = $138
His allowance was $138.

17.

Stamps

19 + 17 = 36

3 u → 36
1 u → 36 ÷ 3 = 12
4 u → 12 × 4 = 48
There were 48 stamps altogether.

18. $\frac{5}{12} × 3 = 1\frac{1}{4}$ pints
Sandra drank $1\frac{1}{4}$ pints of water.
$3 - 1\frac{1}{4} - \frac{1}{3} = 1\frac{5}{12}$ pints
$1\frac{5}{12}$ pints of water were left in the pitcher.

19.

Cake   Cookies   Children

15 u → $\frac{3}{4}$ kg
1 u → $\frac{1}{20}$ kg
2 u → $\frac{1}{10}$ kg = 100 g

Each child received 100 grams of raisins.

Unit 5   AREA OF A TRIANGLE

1. (4)   2. (1)   3. (3)   4. (2)

5. (1)   6. (2)   7. (2)   8. (2)

9. $9 - 4 = 5$ feet
$10 - 2 = 8$ feet
Area of unshaded portion
$= \frac{1}{2} \times 5 \times 8 = 20$ square feet
Area of shaded portion
$= 10 \times 9 - 20 = \underline{70}$ square feet

10. Height = 3.5 inches
Base = 2 inches
Area of triangle $= \frac{1}{2} \times .5 \times 1 = .25$

11. 11.25   12. 54

13. 24   14. 6

15. Area of a triangular side $= \frac{1}{2} \times 5 \times 4 = 10$ square yards
$4 \times 10 = 40$ square yards
The total area of the four sides of the tent is 40 square yards.

16. Area of sail $= \frac{1}{2} \times 12 \times 3.2 = 19.2$ square yards
$\$118 \times 19.2 = \$2,265.60$
He had to pay \$2,265.60 for the canvas.

Units 1–5 ASSESS YOURSELF 2

1. (4)   2. (4)   3. (3)   4. (1)

5. (3)   6. (4)   7. (2)   8. (4)

9. Factors of 20: 1, 2, 4, 5, 10, 20
Factors of 30: 1, 2, 3, 5, 6, 10, 15, 30
Common factors: <u>1, 2, 5, 10</u>

10. 4, 40   11. $\frac{2}{5}$

12. 40

13. $24 \div 2 = 12$ feet   14. $100 \times \$0.40 = \$40$
$12 - 8 = 4$ feet $\$40 - \$24 = \underline{\$16}$
$8 \times 4 = \underline{32}$ square feet

15.
Store 1   Store 2
13 units→ \$26
1 units → $\$26 \div 13 = \$2$
24 units → $\$2 \times 24 = \underline{\$48}$

16.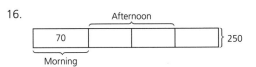
Afternoon
70   } 250
Morning

3 units → $250 - 70 = 180$
1 units → $180 \div 3 = 60$
She had 60 chicken pot pies left.

17. Notes
\$10   \$5 $18 + 32 = 50$

2 units → 50
1 units → $50 \div 2 = 25$
\$10 bills → 25
\$5 bills   → 50
\$20 bills → 1
\$1 bills   → 48
$25 \times \$10 + 50 \times \$5 + 18 \times \$2 + 32 \times \$1$
$= \$568$
The total value of the dollar bills was \$568.

18. $\$4,265 - \$350 = \$3,915$
$\$3,915 \div 15 = \$261$
The amount of each installment was \$261.

19. Total area of living room and 5 bedrooms
$= 30 \times 20 + 5 \times 235$
$= 1,775$ square feet
$\$12 \times 1,775 = \$21,300$
The carpeting will cost \$21,300.

Unit 6 RATIO

1. (3)   2. (3)   3. (1)   4. (2)

5. (2)   6. (3)   7. (2)   8. (3)

9. 4 : 3 : 2   10. 4

11. 3 : 10

12. Amount of unpaid bills
$= \$448 - \$256 = \$192$
$256 : 192 = 4 : 3$

13. 2 : 3

14. $3 - 2 = 1$ unit
1 units → 240 centimeters
$3 + 2 = 5$ units
5 units → $240 \times 5 = 1,200$ centimeters $= \underline{1.2}$ meters

15. 24, 12

16. $\$18 \div 2 = \$9$
Mary has \$9.
$18 : 12 : 9 = 6 : 4 : 3$
The ratio is 6 : 4 : 3.

17. Lawrence's salary $= \$720 + \$180 = \$900$
$720 : 900 = 4 : 5$
The ratio is 4 : 5.

18. Perimeter of each figure
$= 2 \times (31 + 10) = 82$ cm
Area of Figure A = 310 cm²
Length of Figure B $= (82 - 5 - 5) \div 2 = 36$ cm
Area of Figure B $= 36 \times 5 = 180$ cm²

D   ↑ 5 cm
  ← 12 cm →
E
5 cm

Length of D = 5 + 12 = 17 cm
Length of E = (82 − 17 − 12 − 5 − 5 − 5) ÷ 2
= 19 cm
Area of D = 5 × 17 = 85 cm²
Area of E = 5 × 19 = 95 cm²
310 : 180 : 180 = 31 : 18 : 18
The ratio is 31 : 18 : 18.

19. Boys
Girls
612 + 136 + 341

11 units → 612 + 136 + 341 =1,089
1 units → 1,089 ÷ 11 = 99
15 units → 99 × 15 = 1,485
There were 1,485 girls at first.

Units 1–6 ASSESS YOURSELF 3

1. (4)   2. (2)   3. (2)   4. (2)

5. (3)   6. (1)   7. (2)   8. (3)

9. Three million one hundred and twelve
thousand and two

10. 14   11. 78

12. 47 × 39 = 1,833 ≈ 1,800

13. $27 ÷ 100 = 27¢   14. 84 × 2 = 168
45 : 27 = 5 : 3   168 ÷ 6 = 28

15. 36

16. $1 - \frac{1}{10} = \frac{9}{10}$

$\frac{9}{10} ÷ 3 = \frac{3}{10}$

Each person received $\frac{3}{10}$ of the pineapple.

17. Before: After:
4 : 7 : 3   1 : 1 : 1 → 3 : 3 : 3
(since the number of
lilies does not change)
4 − 3 = 1 units
7 − 3 = 4 units
1 + 4 = 5 units
5 units → 84 + 21 = 105
1 units → 105 ÷ 5 = 21
3 units → 21 × 3 = 63
She ordered 63 lilies.

18. 6 + 11 = 17 units
11 units → 1,254
1 units → 1,254 ÷ 11 = 114
17 units→ 114 × 17 = 1,938
There are 1,938 students altogether.

19. Width = 1080 ÷ 45 = 24 yards
Perimeter = 2 × (24 + 45) = 138 yards
$1,104 ÷ 138 = $8
The cost of fencing per yard is $8.

Unit 7 DECIMALS

1. (4)   2. (4)   3. (2)   4. (1)

5. (4)   6. (3)   7. (2)   8. (3)

9. (3)

10. 20.058   11. 10.017

12. 3.1, 3.007, 2.9, 2.75

13. 38.01   14. $\frac{3}{40}$

15. 0.065   16. 0.6

17. $\frac{23}{10}$

18. $\frac{10}{1,000} = \frac{1}{100} = 0.01$ centimeter
Each sheet of paper is 0.01 centimeter thick.

19. Mandy   0.99 meter
Dominic $\frac{4}{5}$ meter = 0.8 meter
Joanne $\frac{7}{8}$ meter = 0.875 meter
Gretel   1.1 meter
Order: Gretel, Mandy, Joanne, Dominic

20. Capacity of the pitcher
= 275 × 5
= 1,375 milliliter = 1.375 liters
370 ml = 0.37 liters
1.375 − 0.37 = 1.005 liters
The pitcher can hold another 1.005 liters of
water.

21. 85 centimeters = 0.85 meters
172 centimeters = 1.72 meters
65 centimeters = 0.65 meters
0.85 + 1.72 + 0.65 = 3.22 meters
The total length of the strips of wood is
3.22 meters.

Unit 8 THE FOUR OPERATIONS
OF DECIMALS

1. (1)   2. (2)   3. (4)   4. (3)

5. (4)   6. (2)   7. (1)   8. (2)

9. 4.74   10. 143.64

11. 1.89   12. 2,800

13. 71.5   14. 33.75

15. 125

16. Amount he spent
= $5.65 × 15 + $4.30 × 12 + $7.40 × 6
= $180.75
$180.75 + $19.25 = $200
He brought along $200.

17. Cost of Type A fencing
= $253 × (2.8 + 3.2 + 4.1) = $2,555.30
Cost of Type B fencing
= $147 × (5.6 + 4.3) = $1,455.30
$2,555.30 + $1,455.30 = $4,010.60
It will cost him $4,010.60 to fence up the ranch.

18. $5.75 × 21 = $120.75
He sold the trout for $120.75.
$498.75 – $120.75 = $378
$378 ÷ 27 = $14
He charged $14 for 1 pound of prawns.

19. Number of cups of sugar used for 1 cake
= 1.5 + $\frac{1}{2}$ × 1.5 = 2.25
Number of cups of sugar needed for 6 cakes
= 2.25 × 6 = 13.5
13.5 – 1.75 = 11.75
She had 11.75 cups of sugar.

20.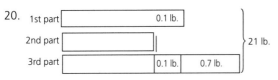

3 units → 21 – 0.1 – 0.1 – 0.7 = 20.1 pounds
1 units → 20.1 ÷ 3 = 6.7 pounds
6.7 + 0.1 + 0.7 = 7.5 pounds
The weight of the third part is 7.5 pounds.

Units 1–8   ASSESS YOURSELF 4

1. (1)   2. (1)   3. (4)   4. (2)

5. (3)   6. (1)   7. (4)   8. (1)

9. 0.625   10. 20

11. $1\frac{1}{4}$ m, 1 m 48 cm, 1.6 m, 180 cm

12. $\frac{5}{16}$

13. Total number of markers
= 28 × 20 = 560
560 ÷ 4 = <u>140</u>

14. 0.003

15. Capacity of 1 big bottle = 125 × 2 = 250 milliliters
Total volume of hand soap
= 125 × 8 + 250 × 4 = 2,000 milliliters = <u>2</u> liters

16. 14 × 14 = 196
Length of 1 side of big square = 14 inches
Length of 1 side of small square
= 48 ÷ 4 = 12 inches
14 : 12 = 7 : 6
The ratio is 7 : 6.

17. $4\frac{1}{2}$ yd. = 4.5 yd.
4.5 yd. – 3.6 yd. = 0.9 yd.
0.9 yd. × $4 = $3.60
The cloth she used cost $3.60.

18. 16 – 3 = 13 pencils
Cost of 13 pencils = $8.25 – $2.40 = $5.85
$5.85 ÷ 13 = $0.45
Each pencil costs $0.45.
$2.40 – $0.45 × 3 = $1.05
Each exercise book costs $1.05.

19. Amount of money that they need
= $2,800 – $450 – $725 = $1,625
Length of time that they need
= 1,625 ÷ 195 = $8\frac{1}{3}$ ≈ 9
It will take them another 9 months before they can buy the entire collection.

Unit 9   PERCENTAGE

1. (2)   2. (1)   3. (3)   4. (4)

5. (4)   6. (4)   7. (4)   8. (2)

9. 15   10. 16 minutes

11. 36   12. 25

13. $\frac{12}{25}$

14. Amount she saves = $10 × 0.2 = $2
$10 ÷ $2 = <u>5 weeks</u>

15. 20

16. Problems solved correctly = 24 – 6 = 18
$\frac{18}{24}$ × 100% = 12%
75% of the problems were solved correctly.

17. (a) Sale price of the dress
= $80 × 0.7 = $56
$56 – $50 = $6
She needs another $6.
(b) $\frac{6}{56}$ × 100% ≈ 11%
She is short of 11% of the money that she needs.

18. 100% – 15% = 85%
85% → 38.25 kilograms
100% → $\frac{38.25}{85}$ × 100 = 45 kilograms
Her weight was 45 kilograms three months ago.

19. Total cost = $1,980 + $1,980 × 0.15 = $2,277
$2,277 ÷ 18 = $126.50
He paid $126.50 each month.

Unit 10   AVERAGE

1. (4)   2. (1)   3. (3)   4. (4)

5. Volume of water Pail B and Pail C can hold together = 2.7 + 0.3 = 3 gallons
(2.7 + 3) ÷ 3 = <u>1.9</u> gallons

6. 37 points   7. 8.7   8. 3.65   9. 20.6

10. Total weight of 6 books = 1.75 × 6 = 10.5 pounds
Total weight of 5 books = 10.5 − 1.9 = <u>8.6</u> pounds

11. Total weight of 2 girls
= 29 kilograms, 850 grams × 2 = 59 kilograms, 700 grams
Total weight of 3 boys
= 32 kilograms, 600 grams × 3 = 97 kilograms, 800 grams
59 kilograms, 700 grams + 97 kilograms, 800 grams= 157 kilograms, 500 grams
Their total weight is 157 kilograms, 500 grams.

12. Total age of 4 students
= 7 years, 2 months × 4 = 28 years, 8 months
Total age of 4 students
= 6 years, 9 months × 3 = 20 years, 3 months
28 years, 8 months − 20 years, 3 months
= 8 years, 5 months
The fourth student is 8 years, 5 months old.

13. Total height of 4 students = 1.75 × 4 = 7 meters
Total height of the other 13 students
= 23.25 − 7 = 16.25 meters
16.25 ÷ 13 = 1.25 meters = 1 meter, 25 centimeters
The average height of the other 13 students is 1 meter, 25 centimeters.

14. Total weight of 2 packages
= 3.725 × 2 = 7.45 kilograms
Total weight of 3 packages
= 7.45 + 2.846 = 10.296 kilograms
10.296 ÷ 3 = 3.432 kilograms = 3 kilograms, 432 grams
The average weight of the 3 packages is 3 kilograms, 432 grams.

Units 1–10   ASSESS YOURSELF 5

1. (3)   2. (2)   3. (3)   4. (4)

5. (2)   6. (1)   7. (2)   8. (4)

9. 8,034,000

10. 73.14 ÷ 7 = 10.448... ≈ 10.45

11. 3.045   12. 600

13. $\frac{3}{2}$, $1\frac{1}{4}$, $\frac{5}{6}$, $\frac{2}{3}$ 14. $4\frac{5}{12}$

15. 13 + 2 + 2 = 17 cm
8 + 2 + 2 = 12 cm
Total area of the picture and margin
= 17 × 12 = 204 cm²
Area of the picture = 13 × 8 = 104 cm²
Area of the margin = 204 − 104 = <u>100</u> cm²

16. Weight of Bag B = 9 + 7 = 16 kilograms
Weight of Bag A = 16 × 2 = 32 kilograms
Total weight of the 3 bags
= 9 + 16 + 32 = 57 kilograms
57 ÷ 3 = 19 kilograms
The average weight of the three bags is 19 kilograms.

17. Discount = $240 × 0.3 = $72
72 ÷ 12 = 6
She could buy 6 scarves with the money saved.

18. (a)   3 units → $90
        1 units → $90 ÷ 3 = $30
        2 + 3 + 5 = 10 units
        10 units → $30 × 10 = $300
        The sum of money was $300.
    (b)   2 units → $30 × 2 = $60
        $\frac{60}{300}$ × 100% = 20%
        Henry received 20% of the sum of money.

19. Total cost = $1.25 × 15 + $2.75 = $21.50
$50 − $21.50 = $28.50
He would receive $28.50 in change.

Units 1–10   ASSESS YOURSELF 6

1. (4)   2. (3)   3. (1)   4. (2)
5. (3)   6. (2)   7. (3)   8. (1)

9. Total area of the faces
= 4 × 8 × 2 + 4 × 3 × 2 + 8 × 3 × 2
= 64 + 24 + 48 = <u>136</u> square centimeters

10. $2\frac{5}{12}$ 11. 3.77

12. $\frac{1}{6}$ × $684 = $114
$114 × 12 = <u>$1,368</u>

13. 12 ÷ 4 = 3
$2.80 × 4 = <u>$11.20</u>

14. 12   15. 7

16. 3 + 2 + 1 = 6 units
1 unit → $78
6 unit → $78 × 6 = $468
The sum of money is $468.

17. $\frac{$150}{100}$ × 110 = $165
$\frac{$165}{100}$ × 107 = $176.55
He spent $176.55 altogether.

18. Duration from 9 a.m. to 8:30 p.m.
= 11 hours, 30 minutes ≈ 12 hours
$3.20 + $1.45 × 11 = $19.15
He had to pay $19.15 in all.

19. $39.58 – $1.90 × 3 = $33.88
Beef → 2 units
Green beans → 1 units
12 + 2 = 14 units
14 units → $33.88
2 units → $33.88 ÷ 7 = $4.84
The can of beef cost $4.84.

Unit 11   ANGLES, TRIANGLES, AND 4-SIDED FIGURES

1. (3)   2. (2)   3. (3)   4. (3)

5. (2)   6. (2)   7. (3)   8. (3)

9. (4)   10. (1)   11. (3)

12. 30   13. 3

14. 140   15. 75

16. ∠BAC = 35°
∠p = 180° – 90° – 35° = <u>55°</u>

17. ∠x = 180° – 60° = 120°
∠y = 60° + 60° = 120°
∠x + ∠y = 120° + 120° = <u>240°</u>

18. 140

19. ∠ACB = 30° + 40° = 70°
∠n = 180° – 70° = <u>110°</u>

20. ∠BCE = 180° – 70° × 2 = 40°
∠DCE = 180° – 70° = 110°
∠DCB = 110° – 40° = 70°

21. ∠CBD = (180° – 50°) ÷ 2 = 65°
∠ABC = 180° – 50° = 130°
∠DBA = 130° – 65° = 65°

22. ∠ABD = 180° – 90° – 65° = 25°
ΔABD and ΔBEC are similar.
∠x = 25°

23.
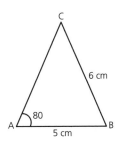

24.
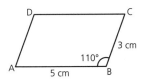
∠BCD = 70°

25.
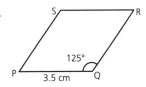
∠SPQ = 55°

Unit 12   VOLUME

1. (2)   2. (1)   3. (4)   4. (4)

5. (3)   6. (4)   7. (1)   8. (1)

9. E   10. A, F

11. 6   12. I

13. B   14. 18

15. 4

16
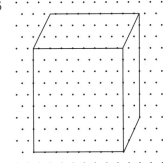

17. $\frac{1}{2}$ × 3 × 2 × 4 = 12 cubic feet
12 – 2.7 = 9.3 cubic feet
9.3 cubic feet of water must be poured into
the tank to make it half full.

18. 20 in. – 16 in. = 4 in.
45 in. × 48 in. × 4 in. = 8,640 cubic inches
8,640 cubic inches ÷ 36 = 240 cubic inches
The capacity of each jar is 240 cubic inches

19. 42 × 28 × 45 = 52 920 cubic centimeters =
52.92 liters
52.92 ÷ 2 = 26.46 ≈ 27
27 bottles are used.

Units 1–12   TEST YOURSELF 1

1. (4)   2. (1)   3. (2)   4. (3)

5. (1)   6. (3)   7. (1)   8. (2)

9. (2)   10. (2)   11. (4)   12. (3)

13. (3)   14. (1)   15. (2)

16. 1 and 3   17. 12, 24

18. 60   19. 54, 63, 72, 81

20. 6 × 6 = 36 square feet
    36 ÷ 4 = <u>9</u> feet

21. 10 ÷ 2 = 5 feet
    2 × (10 + 5) = <u>30</u> feet

22. (48 − 2 × 13) ÷ 2 = 11 inches
    13 × 11 = <u>143</u> square inches

23. 20 × 20 = 400 square inches
    400 × 20 = <u>8,000</u> square inches

24. Monday

25. Tuesday and Thursday

26. 190   27. 50

28. Friday   29. Wednesday

30. Friday

31. 8   32. $20

33. 9 houses   34. M$25

35. S$7   36. 500 milliliters

37. $\frac{1}{2}$ hour   38. $3\frac{1}{2}$ hours

39. 5 × 3 × $24 = $360
    He has to pay $360.

40. Area of picture = 15 × 10 = 150 square inches
    Area of poster board = 17 × 12 = 204 square inches
    204 − 150 = 54 square inches
    The area of the poster board not covered by the picture is 54 square inches.

41. Width = 72 ÷ 9 = 8 yards
    Perimeter = 2 × (9 + 8) = 34 yards
    34 × $6 = $204
    It will cost $204.

UNITS 1–12   TEST YOURSELF 2
1. (4)   2. (1)   3. (4)   4. (3)

5. (3)   6. (2)   7. (2)   8. (2)

9. (3)   10. (1)   11. (4)   12. (4)

13. (1)   14. (4)   15. (1)

16. 2 hours, 20 minutes − 1 hour = 1 hour, 20 minutes ≈ 1 hour, 30 minutes
    $2 + $0.80 × 3 = $<u>4.40</u>

17. $2\frac{1}{2}$ 18. 15

19. $\frac{1}{8}$ 20. 29,000

21. 18 bundles   22. 121.5

23. 180° − 115° = 65°
    180° − 65° × 2 = 50°
    ∠x = <u>50°</u>

24. 29.40   25. 4

26. 228,643   27. 240 cubes

28. 540 boys   29. 59.2

30. Weight of Package B = 3.6 − 1.2 = 2.4 kilograms
    36 : 24 = <u>3 : 2</u>

31. 350 ÷ 7 = 50 ounces
    The weight of the hazelnuts in a bag was 50 ounces.
    274 × 50 = 13,700 ounces
    The total weight of hazelnuts ordered by the customer was 13,700 ounces.
    43,962 + 13,700 = 57,662
    She had 57,662 ounces of hazelnuts at first.

32. Perimeter of the plot of land
    = 2 × (18 + 15) = 66 yards
    $1,254 ÷ 66 = $19
    The fencing cost $19 per yard.

33. 1 kilogram, 8 gram = 1.008 kilogram
    475 grams = 0.475 kilogram
    1.008 + 0.475 + 0.25 = 1.733 kilograms
    Their total weight is 1.733 kilogram.

34. 273 × 396 = 108,108 ounces
    108,108 ounces of mixed nuts was needed.

35. 6 + 5 + 3 = 14 units
    6 + 3 = 9 units
    14 units → 2,030 pounds
    1 units → 2,030 ÷ 14 = 145 pounds
    9 units → 145 × 9 = 1,305 pounds
    The total weight of Crate A and Crate C is 1,305 ounces.

36. 3 meters = 300 centimeters = 15 × 20 centimeters
    2 meters = 200 centimeters = 10 × 20 centimeters
    15 × 10 = 150
    150 tiles are needed.
    150 × $2.30 = $345
    It will cost $345.

37. $\frac{$3,850}{100}$ × 35 = $1,347.50
    He saves $1,347.50 every month.
    $1,347.50 × 12 = $16 170
    He saves $16,170 in a year.

38. 1,100 × 48 = 52,800
    52,800 knives are needed.
    52,800 − 46,455 = 6,345
    They need to pack another 6,345 knives.

Units 1–12   TEST YOURSELF 3
1. (4)   2. (3)   3. (1)   4. (4)

5. (1)   6. (2)   7. (2)   8. (3)

9. (4)   10. (3)   11. (1)   12. (1)

13. (3)   14. (2)   15. (4)

16. 3,020   17. 42 r 22

18. 97 × 89 = 8633 ≈ 8,600

19. $7 + 0.89 + \frac{3}{8} + 1\frac{1}{4} + \frac{9}{5}$
    = 7 + 0.89 + 0.375 + 1.25 + 1.8
    = 11.315

20. $1\frac{1}{4}$ hours = 1 hour, 15 minutes
    1 hour, 15 minutes after 11:50 a.m. = 1:05 p.m.

21. 3 units → 513
    1 units → 513 ÷ 3 = 171
    2 units → 171 × 2 = 342 books

22. 312.62   23. 5 : 3 : 2

24. $1\frac{19}{24}$

25.

26. $1\frac{3}{5}$ 27. 1,020 words

28. 2,000 tiles   29. 6 cars

30. 20 cars

31. 1,426 × 25 + 23 = 35,673
    The number is 35,673.

32. $65.25 × 14 = $913.50
    It cost $913.50 to rent a car for 2 weeks.
    $913.50 − $47 = $866.50
    He had to pay $866.50.

33. 14 − 4.75 × 2 = 4.5 yards
    4.5 ÷ 2 = 2.25 yards
    Its width is 2.25 yards.
    4.75 − 2.25 = 2.5 yards
    Its width is 2.5 yards shorter than its length.

34. $882 + $65 × 6 = $1,272
    He had $1,272 in the bank after 6 months.
    $(\frac{1}{12} × $1,272) ÷ 2 = 53
    Each vase cost $53.

35. $1\frac{3}{4} + \frac{5}{6} + 1\frac{2}{3} = 4\frac{1}{4}$ hours
    She spent $4\frac{1}{4}$ hours on these activities.

36. 2.3 − 0.8 = 1.5 meters
    The length of the remaining piece of rod is 1.5 meters.
    $\frac{1.5}{15} × 100% = 10%$
    It is 10% of 15 meters.

37. Total age of 12 children
    = 4 years, 9 months × 12 = 57 years
    Total age of 8 children
    = 5 years, 10 months × 8 = 46 years, 8 months
    57 years − 46 years, 8 months
    = 10 years 4 months
    10 years 4 months ÷ 4 = 2 years, 7 months
    The average age of the remaining 4 children is 2 years, 7 months.

38. 1,344 ÷ 24 = 56
    There are 56 bags.
    $\frac{56}{100} × 75 = 42$
    She sold 42 bags.
    42 × $0.85 = $35.70
    She would receive $35.70.

Units 1–12   TEST YOURSELF 4

1. (1)   2. (3)   3. (3)   4. (4)

5. (1)   6. (2)   7. (2)   8. (4)

9. (4)   10. (1)   11. (3)   12. (2)

13. (1)   14. (2)   15. (3)

16. 2,450   17. $\frac{9}{4}$

18. 13.43   19. 0.875

20. 28,730   21. 76 − 29 = 47
                             47 × 29 = 1,363

22. 9 × 8 = 72
    864 ÷ 72 = 12 inches

23. 1 minute = 60 seconds
    Number of buttons it can produce in 1 minute
    $= \frac{60}{5} × 4 = 48$ buttons

24. 3 × $0.50 + 2 × $0.20 + 2 × $0.10
    + 6 × $0.05 = $2.40 = 240¢
    240 ÷ 15 = 16

25. 25

26. ∠c = 120° − 40° = 80°
    ∠p = 180° − 80° = 100°

27. 0.049   28. 144

29. January and April   30. 50 stamps

31. $2\frac{1}{3} + 1\frac{1}{2} + 1\frac{5}{6} = 5\frac{2}{3}$ pounds

The total weight of the 3 melons is $5\frac{2}{3}$ pounds.

32. $\frac{3}{12} = \frac{1}{4}$

$\frac{1}{4} \times 12{,}468 = 3{,}117$

It ships out 3,117 cars every 3 months.

33. Total sale = $129 \times \$5 + 210 \times \$2 = \$1{,}065$

$\frac{\$1{,}065}{100} \times 12 = \$127.80$

His commission was \$127.80.

34. Total distance traveled =

$5\frac{3}{5} + 3\frac{7}{10} = 9\frac{3}{10}$ miles

$10 - 9\frac{3}{10} = \frac{7}{10}$ miles

It was short by $\frac{7}{10}$ miles.

35. Total number of cans = $24 \times 4 = 96$
$\$1.25 \times 96 = \$120$
The cost of 4 cartons of sardines is \$120.

36. Average weight of Sofia and Mary

$= \dfrac{26\frac{1}{2} + 29\frac{3}{4}}{2} = 28\frac{1}{8}$ kilograms

5 kilograms, 80 grams = $5\frac{2}{25}$ kilograms

$28\frac{1}{8} - 5\frac{2}{25} = 23\frac{9}{200} = 23$ kilograms, 45 grams

The weight of Perla is 23 kilograms, 45 grams.

37. $30 + 2 = 32$ gallons
$32 \div 2 = 16$
$\$2.45 \times 16 = \$39.20$
He will get \$39.20.

38. $12 - \frac{12}{100} \times 45 - 3.254 = 3.346$ gallons

3.346 gallons of orange juice were left.

Units 1–12   TEST YOURSELF 5

1. (1)   2. (2)   3. (2)   4. (2)

5. (3)   6. (3)   7. (4)   8. (1)

9. (1)   10. (3)   11. (2)   12. (3)

13. (4)   14. (4)   15. (2)

16. 160   17. $\frac{3}{8}$

18. 2,000   19. 0.006

20. 186   21. 7

22. 62,000, 60,000

23. Area of unshaded portion
$= \frac{1}{2} \times 12 \times (8 - 5) + \frac{1}{2} \times 8 \times (12 - 9)$
$= 30$ cm²
Area of shaded portion
$= 12 \times 8 - 30 = 66$ cm²

24. 4   25. $\frac{1}{3} \times 153$ lb. = 51 lb.

153 lb. − 51 lb. = 102 lb.

26. 60   27.   15.5

28. $13 + 8 = 21$ units
21 u $\rightarrow 390 - 8 \times 2 = 374$
1 u $\rightarrow 374 \div 21 = 17$ R $\underline{17}$

29. $1 - \frac{1}{8} = \frac{7}{8}$

$\frac{7}{8} \times 4 \times 3 \times 2 = \underline{21}$ cubic feet

30. 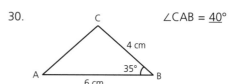 $\angle CAB = \underline{40}°$

31. $2\frac{3}{4} + 1\frac{5}{8} + 2\frac{1}{2} = 6\frac{7}{8}$ miles

The total distance that she covered was $6\frac{7}{8}$ miles.

32. Length of each hose = $2 - 0.28 = 1.72$ yards
$1.72 \times 25 = 43$ yards
She must buy 43 yards of garden hose.

33. $250 \div 12 = 20\frac{5}{6} \approx 21$
He needs 21 bundles.
$21 \times \$2.75 = \$57.75$
The pencils will cost him \$57.75.

34. $20 - 2\frac{5}{8} - 12\frac{3}{4} = 4\frac{5}{8}$ quarts

$4\frac{5}{8}$ quarts of water is needed to fill it up.

35. $0.75 \times 11 = 8.25 = 8\frac{1}{4}$ pounds
$8\frac{1}{4} - 4\frac{3}{8} = 3\frac{7}{8}$ pounds

$3\frac{7}{8}$ pounds of wheat flour were left.

36. $3 + 7 + 8 = 18$ units
$3 + 7 = 10$ units
$10 \div 2 = 5$ units
$5 - 3 = 2$ units
18 units $\rightarrow 2{,}466$
1 units $\rightarrow 2{,}466 \div 18 = 137$
2 units $\rightarrow 137 \times 2 = 274$
Kelly should give Alice 274 marbles.

37. $1\frac{3}{5} + \frac{2}{3} = 2\frac{4}{15}$ hr. = 2 hours, 16 minutes

2 hours, 16 minutes before 7:25 p.m. = 5:09 p.m.

She started working on her math homework at 5:09 p.m.

38. $\$1{,}350 - \$290 = \$1{,}060$
$1060 \div 53 = 20$ months = $1\frac{2}{3}$ years

He will take $1\frac{2}{3}$ years to complete the full payment.

Units 1–12   TEST YOURSELF 6

1. (3)   2. (2)   3. (3)   4. (1)

5. (4)   6. (4)   7. (3)   8. (3)

9. (2)   10. (2)   11. (1)   12. (3)

13. (1)   14. (2)   15. (4)

16. 377,999   17. 287,000

18. 0.19   19. 0.6

20. $4\frac{4}{15}$ 21. 430

22. $\frac{9}{8}$ 23. 105

24. 52   25. 12

26. $40 \div 5 = 8$
$\$0.70 \times 8 = \underline{\$5.60}$

27. 28   28. 1,820

29. $\frac{1}{4}$ hour = 15 minutes
$55 + 15 = 70$ minutes = 1 hour, 10 minutes
1 hour, 10 minutes before 12:30 p.m. =
<u>11:20 a.m.</u>

30. $16 \times 32 = 512$
$1 - \frac{11}{16} = \frac{5}{16}$
$\frac{5}{16} \times 512 = \underline{160}$

31. $1\frac{1}{6} + 4\frac{5}{6} + 2\frac{5}{8} = 8\frac{5}{8}$ feet
The total length of the poles is $8\frac{5}{8}$ feet.

32. Total weight of the curry
$= 4.24 + 1.57 + 0.08 + 8.42 = 14.31$
kilograms
$14.31 \div 45 = 0.318$ kilograms = 318 grams
He estimated that each person would eat
318 grams of curry.

33. $\$180 - \$23.50 - \$18.50 = \$138$
$\frac{\$138}{100} \times 25 = \$34.50$
He spent $34.50 on some books.
$(\$138 - \$34.50) \div 15 = \$6.90$
He spent $6.90 on each child.

34. $1\frac{1}{2} - \frac{3}{4} = \frac{3}{4}$ gallons
$1 - \frac{1}{3} = \frac{2}{3}$
$\frac{2}{3} \times \frac{3}{4} = \frac{1}{2}$ gallons
He had $\frac{1}{2}$ gallons of paint left.

35. $20 - 8 = 12$ centimeters
$40 \times 30 \times 12 = 14,400$ cubic centimeters
The volume of the piece of metal is
14,400 cubic centimeters.

36. $\$50,000 \times \frac{7}{100} \times 8 = \$28,000$
She will withdraw $28,000 worth of interest
after 8 years.

37. $3 + 7 + 4 = 14$ units
$7 - 4 = 3$ units
14 units → 1,428
1 units → $1,428 \div 14 = 102$
3 units → $102 \times 3 = 306$
There are 306 more art books than music
books.

38. $\$0.80 + \$1.20 + \$0.60 = \$2.60$
Each child pays $2.60 for the field trip.
$45 \times \$2.60 = \$117$
The total cost of the field trip is $117.

Units 1–12   TEST YOURSELF 7

1. (3)   2. (2)   3. (3)   4. (1)

5. (2)   6. (2)   7. (2)   8. (3)

9. (1)   10. (2)   11. (1)   12. (1)

13. (4)   14. (2)   15. (4)

16. One million   17. 56

18. $\frac{3}{4}$ 19. $2\frac{5}{18}$

20. 84   21. 342

22. 19.70   23. 24

24. 1.02   25. 7, 7

26. $10 \times 15 = 150$ liters
$150 - 80 = \underline{70}$ liters
27. 24   28. 8.4

29. $8 \times \$0.05 + 15 \times \$0.10 = \$1.90$
$\$20 - \$1.90 = \underline{\$18.10}$

30. $7.15 - 0.91 = 6.24$ kilograms
$6.24 \div 8 = \underline{0.78}$ kilograms

31. $44.82 \div 18 = 2.49$ liters
$2.49 \times 5 = 12.45$ liters
12.45 liters of water can be emptied in 5
minutes.

32. <u>Before:</u>   <u>After:</u>
15 : 2   2 : 1 = 4 : 2
$15 - 4 = 11$ units
11 units → 1,078
1 unit → $1,078 \div 11 = 98$
2 units → $98 \times 2 = 196$
There were 196 ducks at first.

33. $\$24.50 \times 18 + \$2.30 \times 3 = \$447.90$
She spent a total of $447.90.
$\$52.10 + \$447.90 = \$500$
She had $500 at first.

34. $\frac{\$1725}{100} \times 15 = \$258.75$

   The repairs cost $258.75.
   $258.75 + $1,725 = $1,983.75
   The refrigerator cost $1,983.75.

35. $3\frac{1}{2} - 1\frac{5}{8} = 1\frac{7}{8}$ kilograms = 1,875 grams

   1875 ÷ 15 = 125
   He had 125 nails left.

36. 2,450 ÷ 35 = 70 centimeters
   Its width is 70 centimeters.
   (70 + 35) × 2 = 210 centimeters
   The perimeter is 210 centimeters.

37. 13 units → 1,092
   1 units → 1,092 ÷ 13 = 84
   5 units → 84 × 5 = 420
   There were 420 women at the concert.

38. 3.8 × 18 + 1.9 × 5 = 77.9 meters
   77.9 meters of curtain material will
   be needed.

Units 1–12   TEST YOURSELF 8

1. (3) 2. (3) 3. (2) 4. (4)

5. (4) 6. (1) 7. (3) 8. (1)

9. (3) 10. (1) 11. (4) 12. (2)

13. (1) 14. (2) 15. (1)

16. 11   17. 169

18. $2\frac{1}{24}$

19. 0.88, 0.808, 0.8, 0.08

20. 5, 10, 15, 20

21. 30 × 28 = 840 cm²
   11:45 a.m. to 2:45 p.m. = 3 hours
   840 ÷ 3 = 280 cm²

22. 7 × $0.20 = $1.40
   $3.20 − $1.40 = $1.80
   1.8 ÷ 0.3 = 6

23. $\frac{3}{4}$ hours = 45 minutes
   45 − 20 = 25 minutes

24. 425

25. 15 × 12 = 180 m²
   6 × 6 = 36 m²
   180 − 36 = 144 m²
   $\frac{144}{180} \times 100\% = 80\%$

26. 1.45   27. 120

28. 3 : 4   29. 325

30. Thursday

31. 11 units → $19.25
   1 unit → $19.25 ÷ 11 = $1.75
   $19.25 + $1.75 + $6.45 = $27.45
   He had $27.45 at first.

32. $85,400 × $\frac{19}{100}$ = $16,226

   $85,400 − $16,226 = $69,174
   $69,174 × $\frac{35}{100}$ = $24,210.90
   His son received $24,210.90.

33. $7 - \frac{1}{2} - \frac{1}{2} = 6$ meters
   $5 - \frac{1}{2} - \frac{1}{2} = 4$ meters
   6 × 4 = 24 square meters
   24 × $5.99 = $143.76
   It will cost $143.76.

34. 12 × $3.50 + 15 × $0.95 = $56.25
   5 units → $56.25
   1 units → $56.25 ÷ 5 = $11.25
   8 units → $11.25 × 8 = $90
   She had $90 at first.

35. (a) 15 minutes = $\frac{1}{4}$ hour
   $1\frac{3}{4} - \frac{1}{4} = 1\frac{1}{2}$ hour
   She took $1\frac{1}{2}$ hours to complete the paper.

   (b) $1\frac{1}{2}$ hours = 90 minutes
   $1\frac{3}{4}$ hours = 105 minutes
   $\frac{90}{105} = \frac{6}{7}$
   She took $\frac{6}{7}$ of the time limit given to
   complete the paper.

36. Weight of contents in one box
   = $19\frac{3}{5} - 1\frac{17}{20} = 17\frac{3}{4}$ kilograms
   $17\frac{3}{4} \times 6 = 106\frac{1}{2}$ kilograms
   The weight of the contents in 6 such boxes is
   $106\frac{1}{2}$ kilograms.

37. Weight of second bag = 2.842 × 2 = 5.684
   pounds
   Weight of third bag = 5.684 − 0.245 = 5.439
   pounds
   (2.842 + 5.684 + 5.439) ÷ 3 = 4.655 pounds
   The average weight of the 3 bags is 4.655
   pounds.

38. 20 − 6 = 14 centimeters
   28 × 15 × 14 = 5880 cubic centimeters
   The volume of the rock is 5,880 cubic
   centimeters.

Notes

Notes

Notes

Notes